www.wadsworth.com

wadsworth.com is the World Wide Web site for Wadsworth and is your direct source to dozens of online resources.

At *wadsworth.com* you can find out about supplements, demonstration software, and student resources. You can also send email to many of our authors and preview new publications and exciting new technologies.

Plants and People in Ancient Ecuador

The Ethnobotany of the Jama River Valley

Deborah M. Pearsall
University of Missouri

Case Studies in Archaeology: Jeffrey Quilter, Series Editor

Australia • Canada • Mexico • Singapore • Spain
United Kingdom • United States

Anthropology Editor: *Lin Marshall*
Assistant Editor: *Analie Barnett*
Editorial Assistant: *Amanda Santana*
Marketing Manager: *Diane Wenckebach*
Project Manager, Editorial Production: *Catherine Morris*
Print/Media Buyer: *Rebecca Cross*
Permissions Editor: *Sarah Harkrader*

Production Service: *Mary E. Deeg, Buuji, Inc.*
Copy Editor: *Linda Ireland, Buuji, Inc.*
Cover Designer: *Rob Hugel*
Cover Image: *Deborah M. Pearsall*
Text and Cover Printer: *Webcom*
Compositor: *Buuji, Inc.*

The logo for the Archaeology series is based on ancient Middle Eastern and Phoenician symbols for house.

Printed in Canada
2 3 4 5 6 7 06 05

For more information about our products, contact us at:
Thomson Learning Academic Resource Center
1-800-423-0563

For permission to use material from this text, contact us by:
Phone: 1-800-730-2214 **Fax:** 1-800-730-2215
Web: http://www.thomsonrights.com

Library of Congress Control Number: 2002113853

ISBN 0-534-61321-7

Wadsworth/Thomson Learning
10 Davis Drive
Belmont, CA 94002-3098
USA

Asia
Thomson Learning
5 Shenton Way #01-01
UIC Building
Singapore 068808

Australia/New Zealand
Thomson Learning
102 Dodds Street
Southbank, Victoria 3006
Australia

Canada
Nelson
1120 Birchmount Road
Toronto, Ontario M1K 5G4
Canada

Europe/Middle East/Africa
Thomson Learning
High Holborn House
50/51 Bedford Row
London WC1R 4LR
United Kingdom

Latin America
Thomson Learning
Seneca, 53
Colonia Polanco
11560 Mexico D.F.
Mexico

Spain/Portugal
Paraninfo
Calle/Magallanes, 25
28015 Madrid, Spain

To Mike, for all his loving support

Contents

PART III PLANTS AND PEOPLE IN THE JAMA RIVER VALLEY

PART IV A LOOK BEYOND THE JAMA VALLEY

Figures and Tables

FIGURES

TABLES

Foreword

ABOUT THE SERIES

These case studies in archaeology are designed to bring students, in beginning and intermediate courses in archaeology, anthropology, history, and related disciplines, insights into the theory, practice, and results of archaeological investigations. They are written by scholars who have had direct experience in archaeological research, whether in the field, laboratory, or library. The authors are also teachers, and in writing their books they have kept the students who will read them foremost in their minds. These books are intended to present a wide range of archaeological topics as case studies in a form and manner that will be more accessible than writings found in articles or books intended for professional audiences, yet at the same time preserve and present the significance of archaeological investigations for all.

ABOUT THE AUTHOR

Deborah M. Pearsall was born in Detroit, Michigan, in 1950. She grew up in various places in the upper Midwest and graduated from high school in Avon Lake, Ohio, in 1969. She returned to Michigan for college, where she attended the University of Michigan and majored in Anthropology. It was also at Michigan that she became interested in Ethnobotany and studied with Richard I. Ford.

After graduation from college, she enrolled in graduate school at the University of Illinois in Champaign–Urbana and began studying with South American archaeologist Donald W. Lathrap. There she became interested in Ecuador, and participated in Lathrap's excavations at Real Alto, an ancient agricultural village. The study of macroremains and phytoliths from Real Alto became her dissertation research, and she received a Ph.D. in Anthropology in 1979.

In addition to continuing to work in Ecuador, Deborah has conducted research in Peru, Guatemala, Mexico, Puerto Rico, the U.S. Virgin Islands, the Bahamas, Hawaii, Guam, and the midwestern U.S., and has supervised students working in these and other regions. She has taught anthropology at the University of Missouri in Columbia since 1978. She enjoys gardening, especially growing old English roses, and writing. Deborah lives on 80 acres outside Columbia with her husband, Mike DeLoughery.

ABOUT THE CASE STUDY

The story goes that when the inner casket of King Tutankhamun was opened, garlands of flowers were found on the young king. In an instant, however, the colors faded and the flowers shriveled. In a kind of suspended animation of thousands of years, the bouquets from an imperial garden, now long gone, withered in the air. This, perhaps, is one of the most famous moments in the study of ancient plants. For many years, though, bowls of wheat, bags of peanuts, or the root cavities of plants from ancient gardens have informed archaeologists about plant use in the past. For the most part, however, early research on plant remains was serendipitous. Chance discoveries of seeds, flowers, wood, and the like were only noted in cases in which plant remains were well preserved and easily observed. This occurred only rarely, such as in the dry desert conditions of Egypt, Peru, or the Great Basin of the United States or when plants were charred, preserving their structures in the form of durable carbon, or water-logged, protecting them from disintegration in the air.

As Deborah Pearsall notes in the opening pages of this book, the pioneering research of J.G. D. Clark at the site of Starr Carr, in Britain, demonstrated how careful attention to both large and microscopic plant remains can provide great amounts of information about the past. Plants are indicators of both natural conditions and cultural ones. They can tell us about the environment and ecology, for obvious reasons, including information on such phenomena as weather patterns, environmental degradation and the like. As food, medicine, textiles, perfumes, and raw materials for everything from musical instruments to weapons of war, plants can also tell us a tremendous amount of information about culture.

Since the mid-1950s the study of microscopic remains of plants has become increasingly important. This is not only because tiny traces of plants such as pollen can provide additional information to that found by examining larger residues, but also because microremains are sometimes the only traces of plants at archaeological sites. Deborah Pearsall has been one of the leading researchers to exploit the information found in phytoliths. These tiny crystals are made of silica and similar materials that are often preserved when larger plant parts—leaves, petals, and roots—have disintegrated.

This book, *Plants and People in Ancient Ecuador: The Ethnobotany of the Jama River Valley,* is a valuable contribution to the Case Studies in Archaeology series for many reasons. It is a masterful demonstration of how plant remains can be found in difficult conditions of preservation told by one of the leading experts in such techniques. It is also an important contribution to the archaeology of a region, tropical coastal Ecuador, that is poorly known by many, but which deserves much greater attention. In addition, it is a very important contribution to the discussion on the nature of tropical forest agriculture. This is of prime concern not only to archaeologists but also to contemporary people concerned about how to efficiently use the tropical forest for expanding populations without destroying the ecosystem at the same time.

It is thus a great pleasure to present this book to students and their teachers. They will find much to discuss in these pages, from issues of how to recover traces of ancient plants in difficult archaeological conditions, to the interpretation of ancient agricultural systems, to issues of human-environmental relations that were important thousands of years ago and remain important today.

Jeffrey Quilter
Washington, D.C.

Preface

GOALS AND THEMES OF THIS BOOK

Plants and People in Ancient Ecuador: The Ethnobotany of the Jama River Valley explores the interrelationships between the prehistoric residents of a small valley in coastal Ecuador (South America) and the dry tropical forest habitat in which they lived. The book has three related objectives. First, *Plants and People in Ancient Ecuador* is an ethnobotany, a work that explores how, through the medium of culture, people shape and are shaped by the world in which they live. I take as my subject the 3,600-year archaeological record of the Jama River Valley, northern Manabí Province, Ecuador, and determine what plants were selected for food, fuel, building materials, and ritual; evaluate the impact of agricultural activities on the tropical forest environment through interpreting fossil indicators of past vegetation, in this case phytoliths (plant silica bodies); and examine the response of populations to volcanic ash fall disasters.

Second, this book allows me to synthesize results of some 10 years of National Science Foundation–funded research I conducted during the Jama Archaeological-Paleoethnobotanical Project, for which I was coprincipal investigator with Dr. James Zeidler. While I have written book chapters and articles utilizing aspects of my findings, I have yet to synthesize the final results of my analysis of plant remains from 14 archaeological sites and 6 field seasons of observing modern agriculture in the valley. Before Jim Zeidler and I can test models of agricultural evolution, or apply our results to understanding the production of surpluses and the role this played in the evolution of chiefdom-level societies (the larger issues that drove our research), I need to understand the fundamental nature of plant-people interrelationships in the Jama Valley, and model the valley's agricultural potential. I accomplish these goals in *Plants and People in Ancient Ecuador* through an in-depth analysis of the archaeobotanical data from the early Jama-Coaque II phase, the best-preserved data set for the valley, and by analyzing maize and yuca yield data.

Third, *Plants and People in Ancient Ecuador* provides me the opportunity to illustrate the paleoethnobotanical research process, an important component of contemporary interdisciplinary archaeology. I use the Jama case to illustrate how hypothesis-driven research is carried out, and explore the strengths and challenges of interdisciplinary research. I review basic paleoethnobotanical field and laboratory methods, and show how choice of methods influences results. I demonstrate how study of biological materials can contribute to our understanding of "big" issues such as population responses to natural disasters. As a practicing ethnobotanist and archaeologist, I bring to this book my personal perspective on doing research at the natural science "edge" of my discipline.

I have written *Plants and People in Ancient Ecuador: The Ethnobotany of the Jama River Valley* to be accessible in style and content to undergraduate students and beginning graduate students. I also hope the book will be of interest to advanced graduate students and professionals who follow the literature on South American archaeology (especially the tropical lowlands and Ecuador), the evolution of agricultural systems, adaptation to tropical forest environments, paleoethnobotany, and ethnobotany. I have tried to include sufficient detail to satisfy these readers and to address issues of broad relevance in our field.

As the first prehistoric ethnobotany in the Case Studies in Archaeology series, I hope this book illustrates the contribution to archaeology of two major classes of archaeological materials: charred plant remains and phytoliths. I also hope it broadens your vision of what archaeology is, and the types of research in which archaeologists engage, as well as informs you about the archaeology of one of the lesser-known regions of the New World, lowland Ecuador.

ORGANIZATION

The book is divided into four parts. In Part I, What Is Ethnobotany?, I introduce this field, the study of human-plant interrelationships, first by presenting an overview of the field and a few theoretical issues relevant to research in it (Chapter 1), and then by discussing the Jama project, which was designed, in part, as a prehistoric ethnobotany project (Chapter 2). Chapter 2 also includes a description of the Jama-Coaque cultural tradition, the focus of the ethnobotany.

In Part II, Fieldwork, I describe the Jama River Valley and what it was like to live and work there (Chapter 3). I then review the methods used during the project to recover the archaeobotanical data (Chapter 4) and to study modern flora and agriculture in the valley (Chapter 5). Chapter 4 also includes a brief description of the sites tested during the project and establishes the cultural chronology.

Part III, Plants and People in the Jama River Valley, presents the results of my ethnobotanical research, beginning with the identification of all the plant remains recovered during the project, and a discussion of their likely uses (Chapter 6). Discussion of the identified plants continues in Chapter 7, with a focus on plant-people interrelationships during early Jama-Coaque II times. In Chapter 8, I turn to the other major component of my research, the study of maize and yuca productivity, and I present a working model of the reconstructed agricultural landscape. In Chapter 9, I return to one of the central goals of ethnobotany—to understand how humans stay in balance with their environments—and look at changing people-landscape relationships in the Jama River Valley.

Part IV, A Look Beyond the Jama Valley, has two objectives: to place what we learned about the ethnobotany of the Jama River Valley within the larger context of the evolution of tropical forest agriculture (Chapter 10), and to use the Jama case to illustrate how ethnobotany fits into, and contributes to, archaeology (Chapter 11).

ACKNOWLEDGMENTS

I would like to acknowledge the people who contributed to the success of my research in the Jama River Valley and to the preparation of this book. First and foremost, I thank James A. Zeidler for the opportunity to work in the Jama River Valley, and for reviewing the text of this book. I thank my colleagues on the Jama project, Jim, Mary Jo Sutliff, Peter Stahl, Robin Kennedy, Jack Donahue, Cesar Veintimilla, Aurelio Iturralde, and Evan Engwall, for their many contributions to the success of the project. Peter and Robin read and commented on the book text. Andrea Hunter, Marcelle Umlauf, and Eric Hollinger, with Cesar Veintimilla, supervised flotation and assisted with plant collecting and the agricultural study. Noez Zambrano served as my guide during the 1982 field season; Danilo Robles assisted with the ethnobotanical research in later seasons. Mike DeLoughery and George Kennedy helped Robin Kennedy and I during the 1990 season, with Mike returning for the 1991 season. Laboratory work at the University of Missouri was undertaken by Midori Lee, Cesar Veintimilla, Peter Warnock, Russell Gaines, and Karol Chandler-Ezell. Data tables and phytolith diagrams were produced by Brigitte Holt, Justin Nolan, Midori Lee, and William Grimm. Neil Duncan scanned and composed the original figures for this book. Laboratory facilities were provided by the American Archaeology Division, Michael J. O'Brien, Director. Finally, I acknowledge with gratitude the help of numerous people in Jama and San Isidro who allowed me on their land and answered my questions about growing crops in the valley.

1/Ethnobotany
The Study of Human-Plant Interrelationships

> Ethnobotany is a very old discipline. Knowledge of useful plants must go back to the beginning of human existence; even today there is evidence that simian animals seek out plants useful to their purpose.
>
> Schultes and von Reis (1995, 11)

THE FIELD OF ETHNOBOTANY

Ethnobotanists are a diverse lot (Table 1.1). We study the charred leavings of ancient meals to understand past diet, document traditional healing systems of indigenous peoples, record the names of plants recognized by college students, describe the steps in building an outrigger canoe, and analyze the chemical constituents of a healer's remedy.

Ethnobotany attracts practitioners from disciplines as distinct as linguistic anthropology and molecular biology. What unites the field is a common focus on the interactions of plants and people, and the belief that these interactions are important:

> [S]elected cases . . . demonstrate (1) that plants have played a major role in determining the trajectories of modern culture, (2) that the wisdom of indigenous peoples can not only provide insight into the human condition but also enrich Western cultures, and (3) that conservation of plant biodiversity and indigenous plant lore is in the interest of the world community. (Balick and Cox, 1996, 23)

Definitions of ethnobotany have changed over the history of the discipline as research foci shifted, and have diversified in recent years to match the increased diversity of the field. Early definitions emphasized the study of uses of plants by aboriginal/tribal/primitive (i.e., traditional or indigenous) peoples, and this is still at the heart of much ethnobotanical research. Documenting traditional medicines, foods, and objects made from plants before customs and cultures disappeared, with an emphasis on discovering practices and commodities useful to the modern world, prevailed. Later, there was increased interest in how plants were named, and in the system of classification of the natural world revealed by names. The role of traditional cultures in maintaining and altering habitats also became the focus of much

TABLE 1.1 MAJOR AREAS OF MODERN ETHNOBOTANICAL INVESTIGATION*

Field	Main Areas of Investigation
Ethnoecology	Traditional knowledge of plant phenology, adaptations and interactions with other organisms Nature and environmental impact of traditional vegetation management
Traditional agriculture	Traditional knowledge of crop varieties and agricultural resources Nature and environmental impact of crop selection and crop management
Cognitive ethnobotany	Traditional perceptions of the natural world (through the analysis of symbolism in ritual and myth) and their ecological consequences Organization of knowledge systems (through ethnotaxonomic study)
Material culture	Traditional knowledge and use of plants and plant products in art and technology
Traditional phytochemistry	Traditional knowledge and use of plants for plant chemicals (e.g., in pest control and traditional medicine)
Palaeoethnobotany	Past interactions of human populations and plants based on the interpretation of archaeobotanical remains

*From *Ethnobotany: Principles and Applications*, by C. M. Cotton, Table 1.7.

research, with applications to past patterns through archaeology. Plant use in past cultures, always a part of ethnobotany, was explored in the 19th and early 20th centuries by the identification of chance finds, such as desiccated plants from Egyptian tombs and dry caves, and in more recent decades through the systematic recovery of plant remains from archaeological sites. Finally, ethnobotany, in the view of some practitioners, now encompasses folk knowledge of people living in industrialized nation-states (Figure 1.1).

In essense, the name *ethnobotany,* from *ethno* (people, culture) and *botany* (plants), coined by John Harshberger in a lecture in 1895, captures the defining dual nature of the field: ethnobotanists are concerned with people *and* plants. Ethnobotany is interdisciplinary by definition, by practice, and by training. Whether an ethnobotanist comes out of a program in anthropology or botany, he or she must be conversant in both worlds. The definition I prefer (see the title of this chapter) is inclusive and essentially ecological in its focus on *interrelationships* between plants and people.

In the foreword to *Ethnobotany: Evolution of a Discipline,* edited by Schultes and von Reis (1995), Noel Vietmeyer remarks that people of the 20th century began to forget the significance of the plant world to themselves and their history. Manufactured pharmaceuticals replaced herb-derived remedies in many parts of the world, synthetic fibers replaced cotton and wool, and laboratory-based research replaced exploration of the natural world. In recent decades we have seen a reversal of these trends, as society has become concerned with pollution, habitat destruction, loss of biodiversity, dependence on nonrenewable resources, and the disappearance of indigenous peoples and traditional knowledge. One result has been the proliferation of ethnobotanical research and advances in methods; another result has been an increased awareness of the field by the general public.

Figure 1.1 Research subjects of ethnobotany: a. production of a traditional canoe; b. medicinal herbs grown for household use

PALEOETHNOBOTANY

Paleoethnobotany, the study of the interrelationships between people and plants through the archaeological record, has its roots as a scientific discipline in the 19th century (Pearsall, 2000). Dried, waterlogged, and charred pieces of fruits, seeds, fibers, wood, and the like—that is, macroremains—were the first subjects of study,

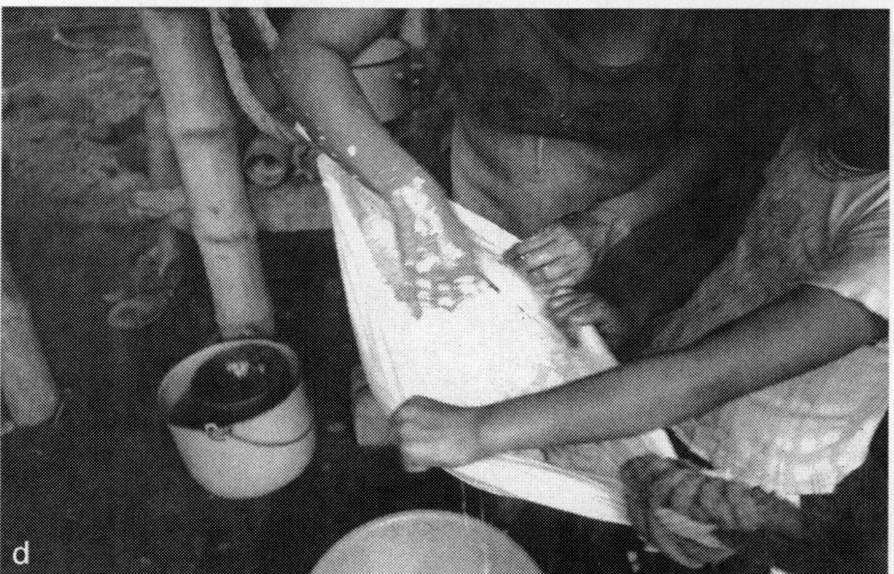

*Figure 1.1 (continued) c. traditional house form in coastal Ecuador; d. processing of yuca (*Manihot esculenta*) for starch*

but pollen and phytoliths (microscopic plant silica bodies) were also discovered and studied systematically in the 19th century. The potential of pollen for identifying plant use and for understanding past vegetation was quickly realized and applied in archaeology, but early phytolith studies were largely ignored until the 1970s, when phytoliths were "rediscovered" by archaeologists.

The history of paleoethnobotany is actually the history of two paleoethnobotanical traditions, one European, and one American. Today these have largely converged in terms of methods and approaches, although there are still differences in research emphasis. Paleoethnobotany began to come into its own in archaeology in the mid-20th century, following publication in 1954 of *Excavations at Star Carr* by British archaeologist J. G. D. Clark. This influential work convinced many of the importance of biological remains for archaeological interpretation. Coupled with increased interest in American archaeology in reconstructing subsistence and paleoenvironment, greater emphasis was placed on the systematic recovery of biological materials in the late 1950s and 1960s. Pollen studies became increasingly common, especially in regions of good organic preservation. Then came the flotation revolution of the 1960s and early 1970s (Figure 1.2).

Flotation utilizes differences in density of organic and inorganic material to separate organic remains from the soil matrix: put a few handfuls of soil in a bucket of water, stir, and rootlets, twigs, and seeds float while pebbles sink (Pearsall, 2000). When properly practiced, flotation allows recovery of *all* size classes of preserved macroremains in quantity, not just plant pieces large enough to be seen during excavation or recovered in the bulk sieve (Figure 1.3). This makes quantitative analysis of macroremains possible. There are many ways to do flotation; common systems include the Illinois Department of Transportation (IDOT) system, a type of manual flotation system used during the Jama project (see Chapter 4), and the Shell Mound Archaeological Project (SMAP) and other water-separator systems, which use pumped water to float botanical materials from the soil. By permitting the systematic recovery of small macroremains, such as seeds, which were previously overlooked in excavations, flotation changed our understanding of past foodways and diet. As archaeologists Frank Hole, Kent Flannery, and James Neely put it:

> Our preliminary report on the 1961 season states confidently that "plant remains were scarce at Ali Kosh." Nothing could be further from the truth. The mound is filled with seeds from top to bottom. All that was "scarce" in 1961 was our ability to find them. (1969, 24).

Today it is not uncommon to find recovery and analysis of pollen, phytoliths, and macroremains written into archaeological research designs, as in the Jama project (see Chapter 2). This is because paleoethnobotanists and archaeologists have found that different types of botanical data produce different kinds of information about past people-plant interrelationships, and have different strengths and weaknesses. Macroremains, for example, are preserved at most archaeological sites, including those tested in the Jama River Valley, by accidental or deliberate charring. Plant foods that are cooked are thus more likely to be preserved as macroremains than those that are not routinely exposed to fire. Further, charred macroremains are not equally robust: a charred fragment of sweet potato is more fragile than a maize cob fragment, which is more fragile than a palm nut fragment. In spite of the interpretative difficulties introduced by differential charring and preservation, macroremains are perhaps the easiest kind of botanical data to link directly to human behaviors such as food preparation. As such, they play a central role in investigating plant-people interrelationships.

Pollen grains and phytoliths do not have to become charred to preserve and, therefore, can document plants that were eaten raw or used in ways that do not

Figure 1.2 Flotation for the recovery of charred plant remains

produce identifiable macroremains. Paleoethnobotanists can learn much about medicinal plant use by analyzing pollen and phytoliths in coprolites (preserved human feces) and latrine soils, for example, since plants used in medicinal teas often transfer pollen or small tissue fragments identifiable by phytoliths to the brew. The outer surface of pollen is composed of sporopollenin, one of the toughest organic materials known. Thus pollen may preserve when other organics do not. Plants used in burial rituals may be identifiable by pollen left after food or other grave offerings decay. Phytoliths, being inorganic, often preserve under conditions in which pollen and macoremains are both destroyed, for instance, in clayey sediments that are habitually wetted and dried (Figure 1.4). As we found in many samples studied during the Jama project, shrinking and swelling of clays can break up charred remains and fracture pollen grains, but leave phytoliths intact.

Pollen and phytolith data are central in research focused on reconstructing human-landscape interrelationships (Pearsall, 2000). Many plants are wind-pollinated, and when the resulting "pollen rain" falls on the surface of a lake or pond, it sinks and becomes incorporated into bottom sediments. Phytoliths carried by streams entering the lake, as part of the silt load from runoff in the watershed, also become incorporated into sediments. By extracting a stratigraphic core from the lake basin (by driving a long tube into the bottom sediments) and identifying the recovered pollen and phytoliths, one can document the vegetation surrounding the lake. Human activities such as clearing forest for agriculture leave their mark on vegetation, and this in turn changes the pollen and phytolith record in the lake. Phytolith and pollen data from lake cores often complement each other: some plants can be identified by their pollen grains, others by their phytoliths.

Figure 1.3 A small seed recovered from an archaeological context by flotation

A FEW THEORETICAL ISSUES

Paleoethnobotany is more than just identifying plants through their preserved pollen grains, macroremains, or phytoliths. Paleoethnobotanists are interested in understanding how people were affected by the plant world—how available plant resources shaped cultural practices, influenced health, set seasonal activities, and determined settlement history—and what influence people had on that world—how subsistence and other activities affected plant distribution, abundance, and population structure, and the impact of subsistence activities on vegetation and landscapes. This is an ecological approach: paleoethnobotanists attempt to understand how parts of a prehistoric ecosystem functioned, how particular human populations articulated with their natural world, and what the consequences were of not achieving a balance of population size and resource base.

Not all paleoethnobotany is about food (e.g., fuel can be identified, and medicinal plant use investigated), but as the food quest is central to the success of a population, understanding the nature of the food quest is central to most paleoethnobotanical studies. By identifying what plants were used as foods and investigating how these foods were procured, paleoethnobotanists can reconstruct prehistoric subsistence systems and gain insight into diet and health practices.

I discuss how plant remains are identified in the first part of Chapter 6. Here I would like to focus on some issues and assumptions related to reconstructing past subsistence, diet, and health. First, as in much of archaeological interpretation, what paleoethnobotanists do is attempt to *model* past practices, in this case, past people-plant interrelationships, from the material record of those practices. If a model is

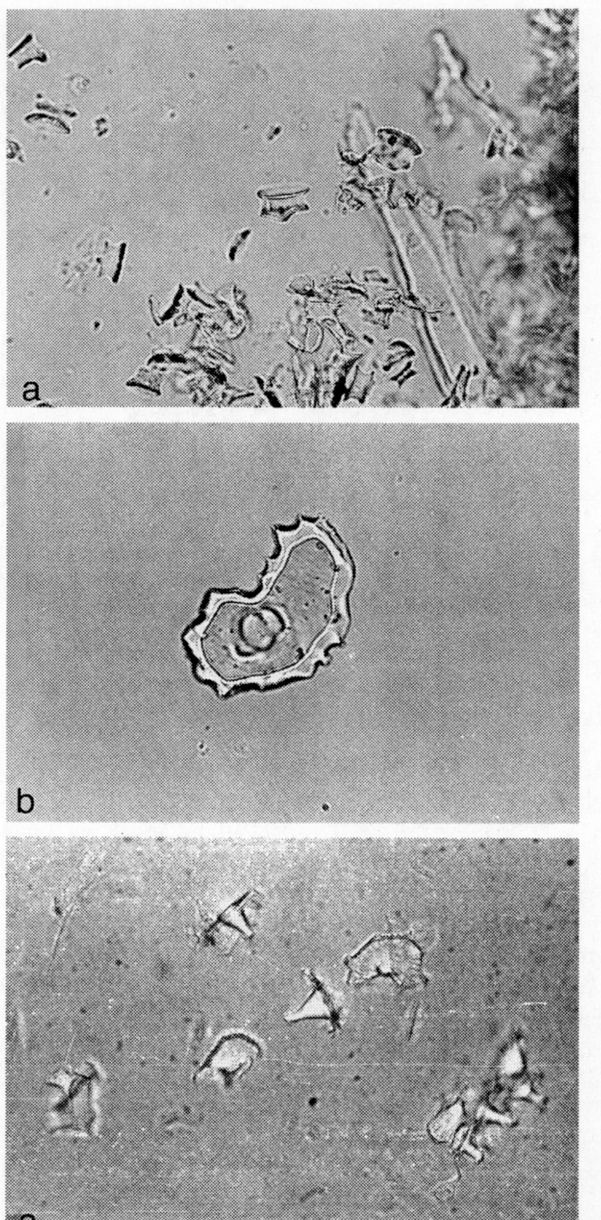

*Figure 1.4 Examples of phytoliths observed in samples analyzed during the Jama project: a. wavy-top rondel from maize (*Zea mays*); b.* Burseraceae *seed phytolith; c. sedge (*Cyperaceae*) seed epidermal phytoliths*

based on several independent lines of evidence, incorporates practices known from living peoples, and makes sense in terms of plant distributions and habitat, then you, the reader, may find the model convincing. This does not mean that the model is right, however. Another combination of practices, some not known today, might have produced the same results.

For example, today some traditional agriculturalists in the lowlands of South America rely on the starchy roots of yuca (or manioc, *Manihot esculenta*) as a mainstay of their diet, and grate those roots into flour to toast as yuca cakes, but that does not mean that those practices were followed in the past. If I find fragments of charred yuca root in flotation samples, I do not know how the roots were processed or whether the yuca was eaten, used in ritual, or grown as an ornamental plant. If I also find grater chips and griddle fragments covered with yuca starch, then I can draw inferences about how the roots were processed, but I still have not proven that the yuca was eaten: perhaps the toasted cakes were used to bait traps or traded to other groups. If I find coprolites stuffed full of yuca root fibers, this makes a strong case that the roots were eaten, but I still do not know how important yuca was in the diet. To make a convincing case that yuca was very important, I would need multiple occurrences of charred roots, grater chips and griddle fragments from many households, bone chemistry data consistent with heavy yuca use, evidence of extensive forest clearance for agricultural fields, or preferably a combination of these lines of evidence.

The reality is that paleoethnobotanists are lucky to get one or two lines of evidence in support of how people interacted with plants, and must rely heavily on ethnographic analogy (how plants are used by traditional peoples living in similar habitats) and an understanding of the plants themselves (growth requirements, geographic distribution, nutritional content, and the like) to develop convincing models. Future research hopefully will provide new evidence to support a model; that is, over time, the weight of evidence from other studies hopefully will give similar results that will lead to general acceptance of a model.

Because paleoethnobotanists rely heavily on ethnographic analogy and observations of present-day distributions and requirements of plants, it is important to consider the limitations and strengths of these sources. Study of living traditional peoples can tell us much about how people gather and grow plant foods. Traditionally, however, gathering and farming have been considered distinctive modes of interacting with plants. But there are, in fact, plant-people interactions that do not fall neatly into these modes: tending/encouraging wild species, transplanting wild species, gathering weedy plants encouraged by human disturbance, growing plants that exhibit different degrees of genetic change, gathering plants escaped from cultivation, and using "starvation" foods in a bad year. Further, subsistence farmers around the world engage in plant-gathering activities on a seasonal basis, while "classic" hunter-gatherers like the !Kung engage in herding and trade for agricultural produce with their neighbors. The complexity of plant-people interrelationships of traditional peoples in the modern world may have analogues in prehistory.

The same plant may be the object of different kinds of interactions in the ethnobotanies that paleoethnobotanists consult. Palms (Arecaceae), for example, are very useful plants of the lowland tropics. Palm fruits and fronds may be gathered from palms growing in natural stands, from planted palms, or from individuals left standing when trees are cut to make fields. (Palms and other useful trees may be visited for years after a field is left fallow, i.e., rested.) Finding charred palm fruits or palm phytoliths does not tell us which patterns were followed.

Further, ancient interrelationships, like those between people and palms in the tropical lowlands, may lead to changes in plant distributions and densities, making it challenging to use modern data to model resource availability. It can be argued, for example, that large tracks of forest in the Amazon basin of South America are anthropogenic, the result of millennia of activities that encouraged some wild plants and led to the decline of others. "Wild" stands of useful trees thus may not reflect "natural" distributions and stand densities of species at all, but may be the product of ancient interactions with people. Vegetation may also become so altered by modern land-use practices that it is next to impossible to model past patterns from the present-day plant cover. Coring in sedimentary deposits can reveal the history of vegetation change in a region, including the spread of anthropogenic plant communities and the presence of resources locally extinct today.

Sometimes the plants themselves narrow the range of possible ethnobotanical interactions. This is often the case with domesticated plants (but is not limited to them), especially those that are fully domesticated, that is, plants that have lost the ability to reproduce without the intervention of people. Maize is a good example of such a plant. Finding maize in archaeological sites in the Jama River Valley of coastal Ecuador tells me a number of things about people-plant interrelationships. Maize does not occur wild in South America, and it cannot self-seed successfully (i.e., it cannot escape from cultivation): it has to be planted, harvested, and replanted. So finding maize remains in Jama samples tells me that people prepared garden beds, planted maize seeds, cultivated the plants (at least minimally), harvested the maize ears, and stored seed for future planting. Since these activities must be fit in with other subsistence activities, I can begin to develop a model of a seasonal subsistence round that accommodates growing maize. Further, by consulting sources on water and soil requirements of maize, or gathering information on maize yields in the study region (the approach I took in the Jama project), other dimensions can be added to the model: where the best maize lands are located, and how much maize can be produced using traditional cultivation techniques.

Another issue related to the use of ethnographic analogy is the problem of how to interpret pollen, macroremains, or phytoliths from plants for which there are no ethnobotanical data. It may be clear (to the paleoethnobotanist) that a plant was used or even grown (e.g., remains are ubiquitous in samples, there are storage caches, remains are altered from wild forms, occurrence is outside the native range of the species), but how can this be demonstrated if the species dropped out of use during prehistory, or its historical use was not documented? A classic example is the native North American cultivated seed complex, sunflower, sumpweed, chenopod, erect knotweed, canary grass, barnyard grass, and little barley: only sunflower was still in cultivation when Europeans arrived in North America. The argument for the rest of the complex has been made using patterns of occurrence of the remains themselves at numerous Woodland and Mississippian sites.

In summary, the goal of paleoethnobotany is to understand past interrelationships between people and plants through the archaeological remains of those interrelationships. The starting point is the recovery and identification of plant remains from (hopefully) well-understood and dated archaeological contexts; the end result is a model of past people-plant interactions. This is an ecological approach to understanding the past: paleoethnobotanists attempt to determine how parts of a prehistoric ecosystem functioned. To achieve this end, they must research the ecology of the identified plants, be familiar with ethnobotanical studies from their region, understand the nature of the archaeological record, and be aware of the pitfalls (and how to mitigate them) of relying on modern plant distributions and uses. When the laboratory work and interpretations are done, the resulting model of past plant-people interrelationships will contribute to ethnobotany, and be tested by future research.

The final point I would like to touch on is the concept of humans remaining in balance with the natural world, and how paleoethnobotany contributes to our understanding of this process. Some of the liveliest debates in the undergraduate ethnobotany class I teach concern the notion that native peoples traditionally live in harmony with the natural world. Not only do their activities and traditional knowledge help sustain endangered ecosystems and species, as demonstrated by a number of ethnobotanical studies, but they are the logical, rightful guardians of these systems. These are, needless to say, complex issues that can pit ethnobotanists against some conservationists, who feel that people have no place in protected ecosystems. These are also issues that can be informed by looking at the past. Most ethnobotany students are astonished the first time I show them a pollen diagram that documents extensive clearing of a tropical rainforest—by the native peoples who lived there.

In bounded ecosystems, populations must remain at levels below the carrying capacity of their food resources or else increase resource availability (e.g., by growing plants or through technological advances like terracing or irrigation), migrate in search of new resources, or suffer malnutrition and eventual starvation. Growing more crops is usually associated with changes in vegetation (more trees are felled, fields are used longer with less forest regrowth), and sometimes with landscape alteration. Whether, and how, balance between population and resource base was achieved is something paleoethnobotanists can investigate: it is not an assumption we should make. In the case of the Jama Valley, people lived there successfully for some 3,600 years: investigating how they lived in this dry tropical forest setting for so long, and how they responded to periodic disasters such as being blanketed by volcanic ashfall from the highlands, may provide insights into the sustainability of agriculture in the tropical lowlands.

Did the inhabitants of the Jama Valley cut down tropical forest trees? Yes, they did. Did they develop ways of using the same parcels of land over and over again for agriculture in a sustainable manner? Yes, they did. Did their activities lead to localized plant and animal extinctions? Probably: we know, at least, that certain species declined, while others expanded, as habitats were altered. Were the inhabitants of the valley living in balance with the natural world? Yes, if we measure this by cultural survival; no, if we mean that the valley looked the same at A.D. 1500 as it did at 1600 B.C. How do I know the answers to these questions? Read on.

2/The Research Process
An Overview of the Jama Project

WHY THE JAMA VALLEY? A BRIEF HISTORY OF THE PROJECT

Northern Manabí province in coastal Ecuador is one of many "gray areas" on the archaeological map of this small South American country (Figure 2.1). The region is perhaps best known as the center of the Jama-Coaque culture (Zeidler and Pearsall, 1994a). Jama-Coaque is a late prehistoric culture previously assigned to the Regional Developmental period (500 B.C. to A.D. 500), a time in Ecuadorian prehistory characterized by geographically restricted, small-scale chiefdoms (Meggers, 1966). Better known than the archaeology are the elaborate ceramic artifacts looted from sites across the province. Before the start of the Jama project, no systematic archaeological survey had been carried out in the region, no absolute dates for the Jama-Coaque culture existed, and nothing was known about how prehistoric populations made their living in this dry tropical forest region.

James A. Zeidler began working at the prehistoric San Isidro site, in the middle reaches of the Jama Valley, in 1981. The San Isidro site, located beneath the modern town of that name, was well known for the size of its central platform mound, and the beautiful ceramic, shell, and stone artifacts "mined" from it by local antiquity traffickers. That first season, as well as the 1982 and 1983 seasons at San Isidro, were supported by grants to Jim Zeidler, and benefited from logistical support from the Escuela Superior Politécnica del Litoral, Guayaquil (ESPOL), and participation of ESPOL archaeology students in the field research. This early research was focused on establishing a basic cultural chronology for San Isidro, and on understanding the sequence of mound building at the site (Figure 2.2).

From the beginning of research at San Isidro, soil samples were collected for recovering charred plant remains and faunal materials by water flotation, and for extraction of microscopic pollen grains and phytoliths. Through my ties to ESPOL, and prior research with Zeidler at the Real Alto site in coastal Guayas province, I became involved with the research program at San Isidro. Very soon we developed the idea for a joint project on a regional scale.

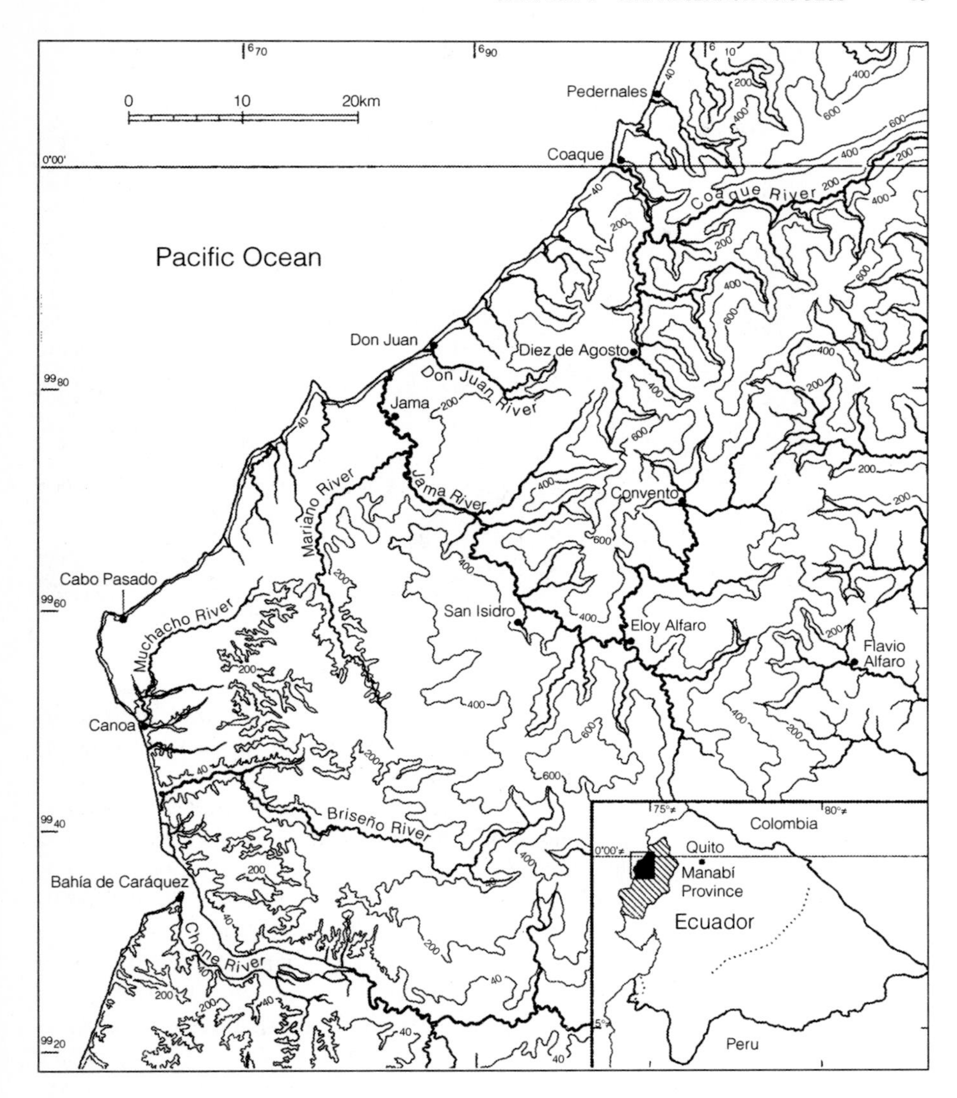

Figure 2.1 General location map for Ecuador, Manabí Province, and the Jama Valley. Figure 1.1 in Regional Archaeology in Northern Manabí, Ecuador, *ed. Zeidler and Pearsall. Used with permission.*

The Jama Valley Archaeological-Paleoethnobotanical Project was initiated in the mid-1980s for the purpose of conducting an interdisciplinary archaeological investigation of the emergence and evolution of complex societies in northern Manabí province. From 1988 on, fieldwork and analysis were supported by grants from the National Science Foundation, with additional funding provided by the Museo Arqueológico of the Banco Central del Ecuador. In 1994 Zeidler and I published an edited volume on our research through 1988 (Zeidler and Pearsall, 1994b),

Figure 2.2 The central mound of the San Isidro site

which focused on regional chronology building, and summarized preliminary environmental and subsistence data. The results of a regional archaeological survey and site testing program, carried out during the 1989, 1990, and 1991 field seasons, and testing of hypotheses regarding the evolution of complex societies and agriculture, will be the subject of a second project volume, currently in preparation.

My first field season in the Jama Valley, October–December 1982, was spent in the San Isidro area with Zeidler and an archaeology field school from ESPOL. Fieldwork was scheduled to take advantage of the late dry season. (It was far from dry, being the start of the 1982–1983 El Niño, but that is another story.) During those three months I began to get to know the region, its people, and the archaeology, and started the fieldwork that would allow me to identify and interpret the archaeobotanical record. I collected plants that were in flower, especially arboreal species, and took surface soil samples. I watched preparation of fields for the late December/early January maize, or corn, planting, and made notes on where maize fields were placed in relation to other uses of land. I measured the yields of yuca, as well as maize grown under irrigation for harvest as *choclos* (green corn, a local delicacy). These were all activities I could accomplish during the late dry season.

The next two ethnobotanical field seasons—March 1983, conducted by Cesar Veintimilla, then an ESPOL student, and March 1988, carried out by myself, Veintimilla, and two then MU graduate students, Marcelle Umlauf and Andrea Hunter—were timed for the end of the rainy season. This allowed us to collect plants flowering then, and to observe the maize harvest. Umlauf returned in summer 1989 to carry out flotation for the project and gather more maize harvest data; then MU graduate student Eric Hollinger did the same in summer 1990. My final field seasons in the Jama Valley were two visits to finish the maize and yuca agricultural study, in January 1990 and August 1991.

ISSUES THAT SHAPED OUR RESEARCH

It is a valuable contribution to regional archaeology to "fill in" an archaeological gray area, and to document prehistoric cultures being destroyed by looting. However, in order for our research in northern Manabí to contribute to archaeology outside Ecuador, Zeidler and I needed to frame the research around issues of general interest in the discipline. In other words, if our research were issue-driven, our results would become a case study useful for comparison to the research of others.

The interrelated processes of agricultural development and the evolution of social and political complexity in the lowlands of the Neotropics are issues of growing importance in New World archaeology. Of particular interest are the roles played by population dynamics, agricultural productivity, and labor intensification in the development of complex, hierarchical societies such as chiefdoms (Zeidler and Pearsall, 1994a). Coastal Ecuador provides an interesting case study for this type of research question, since the first chiefdoms developed early, during the Formative period (3400–355 B.C.), and never attained state-level complexity, like the Inca empire of the central Andes. Rather, late prehistory in coastal Ecuador, as well as much of northern South America and Central America, was characterized by regionally based chiefdoms of varying sizes that interacted with each other for long periods of time.

Out of our early research at San Isidro, then, grew the long-term goal of exploring through research in the Jama Valley how complex sociopolitical organizations emerged in northern Manabí. In particular, we focused on the interplay of settlement dynamics and agricultural production over the regional landscape. Together, settlement and subsistence data can tell us much about how human labor and surplus production were mobilized to support the development of a complex society on the Ecuadorian coast.

Of particular interest to me is investigating the process of agricultural intensification in the lowland tropics (Pearsall, 1995b, 1999; Piperno and Pearsall, 1998). One of my goals for the Jama project was to test models of agricultural evolution and cropping intensification by using biological data recovered from archaeological sites in the valley. One model I hoped to test was the following: that while domesticated plants were used during the Preceramic and early Formative periods on the coast, subsistence was not primarily agricultural, in the sense this term is used by David Rindos (1984) in his coevolutionary model of agricultural origins. In other words, plant-people interrelationships were not primarily focused on activities that affect the environment inhabited by domesticated plants. I proposed that the transformation of plant-people interrelationships to primarily agricultural activities occurred in the late Valdivia times, with the emergence of societies fully dependent on agriculture occurring much later.

Since my first formulation of this model, I have become increasingly interested in the *types* of activities carried out by early agriculturalists, that is, the kinds and sequences of interactions between people and landscapes that resulted in the emergence of an agroecology—a landscape extensively manipulated by people for crop production. I review various models for the evolution of tropical forest cropping practices in Chapter 10.

The Jama Valley should provide an excellent opportunity to test models of the nature of prehistoric agriculture in the dry tropical forest. Occupation of the valley begins in terminal Valdivia times (Piquigua phase in the Jama chronology; see Chapter 3), when the shift to primarily agricultural relationships may begin. Following Rindos, early agriculturalists should have a subsistence system of high diversity, serving to minimize risk. Over time, diversity decreases, and productivity of stable crops increases as the agricultural system evolves. Population, which is held to be a consequence rather than a cause of increased agricultural activity, subsequently increases.

Without going into detail here, I developed a series of predictions of changes in diversity of foods over time to be tested using the botanical data recovered from excavations in the Jama River Valley, as well as predictions concerning site placement and cropping strategies to be tested against the settlement data and agricultural productivity studies (Pearsall, 1999). These research issues, among others, will be addressed in the second project volume.

WHY THIS ETHNOBOTANY?

Before Zeidler and I can test models of agricultural evolution or apply our results to understanding the production of surpluses and the role this played in the evolution of complex societies, there are two intermediate goals to be accomplished. The first is to understand the fundamental nature of plant-people interrelationships in the valley; the second, to develop a template of prehistoric agricultural productivity. These are the goals of this ethnobotany.

I will focus, then, on describing plant-people interrelationships using archaeobotanical data from the early Integration period, or early Jama-Coaque II, and on modeling yields of maize and yuca using the results of the agricultural study. There are two reasons for focusing on early Jama-Coaque II. First, the best-preserved sample of charred botanical remains from the project comes from a bell-shaped pit at the Pechichal site, dated to the first phase of Jama-Coaque II, the Muchique 2 (A.D. 400–750) phase. Overall, the macroremain sample is largest from Muchique 2 and the subsequent Muchique 3 phase (A.D. 750–1250). Occupation at San Isidro, the ceremonial center occupied from initial settlement of the valley, apparently ceases at the end of Muchique 3, at ca. A.D. 1250. Our sample of Muchique 4 (A.D. 1250–1532) is quite small, in the botanical data, and limited to El Acrópolis, a mound site in the coastal zone. Thus a focus on Muchique 2 and 3 incorporates the largest data set, and covers the Jama-Coaque II occupation at the San Isidro site, at which we have earliest evidence for complex sociopolitical organization in the valley.

There is considerable continuity in vessel form and decoration among all Muchique phases, including Muchique 1, the Regional Developmental or Jama-Coaque I period (Zeidler and Sutliff, 1994). Indeed, the Jama-Coaque tradition continued, apparently uninterrupted, from Muchique 1 deposits below a thick volcanic ashfall, or tephra deposit (Tephra III), to a series of later cultural deposits (Muchique 2–4) postdating the eruptive event. By contrast, two earlier tephra events discovered during our research (Tephra I, Tephra II) mark cultural discontinuities in the valley. The second reason for focusing on early Jama-Coaque II, then, is to document the nature of the subsistence system that allowed people to adapt successfully to environmental impacts of the Tephra III event.

THE JAMA-COAQUE CULTURAL TRADITION

At the beginning of the Jama project, there was uncertainty in the identification of late prehistoric cultures occupying northern Manabí, or the region from around Bahía de Caráquez to near the border of Esmeraldas province. The Jama-Coaque tradition was considered by Smithsonian archaeologist Betty Meggers (1966) to be limited to the Regional Developmental period (ca. 500 B.C. to A.D. 500). During the Integration period (ca. A.D. 500–1500), the Manteño culture (centered in coastal Guayas and southern and central Manabí) extended to just north of Bahía de Caráquez, and the rest of northern Manabí was a gray area. Ecuadorian archaeologist Emilio Estrada (1957) maintained, on the basis of test excavations and surface collections at sites along the Manabí coast (including the towns of Jama and Coaque), that the Jama-Coaque tradition extended temporally through the Integration period, and geographically north through Manabí to around the border of Esmeraldas province. Our research in the Jama Valley supports Estrada's finding that Jama-Coaque was the cultural tradition of both the Regional Developmental and Integration periods in this region.

The Regional Developmental period marks the beginning of regional diversification and development in coastal Ecuador, phenomena seen widely in South America at about this time (Bruhns, 1994). From a coastal late Formative (Chorrera) base emerged several distinctive ceramic traditions, including Jama-Coaque. Pottery forms proliferated; new decorative techniques appeared; art styles were elaborated. Some ceremonial centers occupied during the Formative period, such as San Isidro, expanded in size of resident population as well as in the complexity and size of the ceremonial precinct during the Regional Developmental period, while others were abandoned as the loci of political power shifted. These trends became more marked in the subsequent Integration period, which is characterized by growing sociopolitical complexity, occupational specialization, organized trade, and increased evidence for warfare among polities.

There is little that can be generalized about the layout of Jama-Coaque villages and towns, since San Isidro is the only site that has been mapped in any detail, and the Jama project included the first systematic archaeological survey carried out in the region. Further, no domestic or ceremonial structures have been excavated in their entirety. A house-shaped ceramic vessel from Jama illustrated by Meggers shows a rectangular plan with an opening on one short end, a slightly curved ridgeline, and a steeply sloped roof. The vessel is decorated with green, yellow, and red postfired painting.

Ethnohistoric accounts of the late Jama-Coaque II (Campace) polity, gathered between 1526 and 1531, provide some insights into the nature of settlements. Coaque, located just north of the Jama River Valley, was sacked by Pizarro on his third voyage in 1531; gold, silver, emeralds, and cloth were sent back to Panama and Nicaragua. Pizarro's party remained in Coaque for some eight months, awaiting the return of their ships. Zeidler (1987), in his review of five eyewitness accounts from this occupation, notes that the Spaniards found a large, well-constructed town of 300–400 large houses located on both sides of the river. Coaque had exceptionally well-provisioned storehouses of gold and silver ornaments and beads, emeralds, and

Figure 2.3 Large pottery figure from the vicinity of Caráques (Photo plate CIII from Saville, 1910.)

Spondylus shell beads, as well as large quantities of white cotton cloth. Foodstuffs were also stored in abundance; these included maize, fruits, chiles, fish, and cakes, possibly made of maize or yuca. One eyewitness remarked that the storehouses contained enough provisions to provide for the 180 or so Spanish for three or four years. Coaque was likely a major port-of-call of Manteño sea-faring merchants, and perhaps the site of a Manteño trading enclave as well.

Jama-Coaque pottery is diverse and well-made. In addition to vessels and figurines, pottery forms include spindle whorls, headrests, stamps, bird and animal whistles, flutes, low stools with cylindrical legs, and elbow pipes (Meggers, 1966). Figurines are mold-made, often with appliqué, and depict elaborate costumes, headdresses, and body ornamentation. Male, female, zoomorphic, and "monstrous" forms are known (Figure 2.3).

Because few examples of excavated figurines or figurine fragments exist, it is difficult to interpret their cultural significance and function. We do not know, for example, which types of dress and ornamentation depicted on figurines reflect daily life, and which are ceremonial attire, or how dress reflects social status and gender differences. Meggers notes that female figurines are often less elaborately dressed and ornamented than males, but there is variation in dress among both male and female figures. However one interprets them, Jama-Coaque figurines provide a glimpse at rich artistic traditions in cloth, feathers, metalwork, beadwork, and body painting that are otherwise poorly represented in the archaeological record.

As a result of qualitative and quantitative seriation of Jama-Coaque ceramics from test excavations in the Jama Valley, Zeidler and Sutliff (1994) characterized the

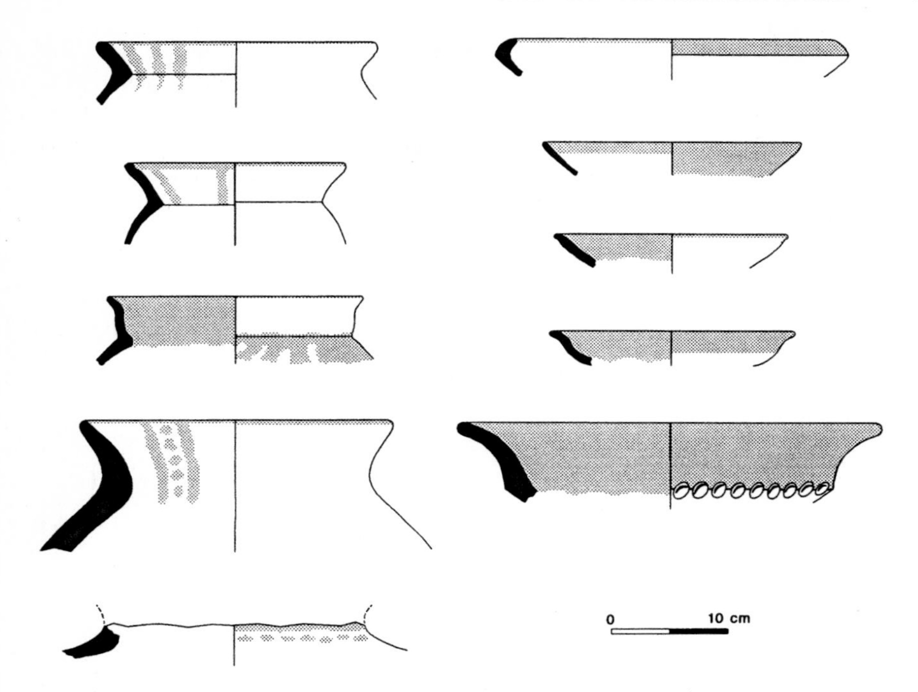

Figure 2.4 Jars characteristic of the Muchique 2 period (left) and bowls from Muchique 3 (right). Figure 7.12 and 7.13 in Regional Archaeology in Northern Manabí, Ecuador, *ed. Zeidler and Pearsall. Used with permission.*

ceramics of four prehistoric phases (Muchique 1–4), and described the diagnostic markers of each phase. Common pottery vessel forms include flat-rimmed polypod bowls with three to five legs (mammiform, cylindrical, conical, and loop shapes), pedestal bowls with bell-shaped bases (*compoteras*), wide-mouthed jars (*ollas*), bowls, and plates (Figure 2.4). Two new *olla* forms appear in Muchique 3: a constricted neck olla or jar, and an open-mouthed carinated "cuspidor." Decorative techniques include red painted bands and dots, lip notching, finger indentation, polished red slip, and fine-line incision. Decorative techniques derived from the late Formative period Tabuchila (Chorrera) tradition that continue into Muchique 1 but then disappear include negative painting and wide-line incision.

• • •

PART I: SUMMARY

In Part I of this book, What Is Ethnobotany?, I have introduced the field, first by providing an overview of the study of human-plant interrelationships, then by presenting the goals and some of the major hypotheses of the Jama Archaeological-Paleoethnobotanical Project, which was designed in part as a prehistoric ethnobotany. As a result of Jim Zeidler's research at the San Isidro site in the early 1980s, we came to understand the cultural chronology of the region, determined that the valley was first settled in the last part of the early Formative period, and identified the late prehistoric cultures of the valley as part of the Jama-Coaque tradition. This set the stage for developing models about how complex societies (chiefdoms) emerged in the region, and about the evolution of the agricultural system that underlay this development.

But how does one get from model building and hypothesis formulation to the data needed to test these ideas? Through fieldwork, of course, which is the subject of Part II.

3/Living and Working in the Jama Valley

DESCRIPTION OF THE REGION AND HISTORY OF SETTLEMENT

Between 1906 and 1908, archaeologist Marshall Saville led three expeditions to explore the antiquities of Manabí province. Part of an ambitious project underwritten by George G. Heye to collect and publish data relating to the pre-Columbian inhabitants of the region lying between Peru and Panama, Saville's descriptions of Manabí not only document the extent of prehistoric occupation of the Manabí coast, but also provide a glimpse of the Jama region in the early decades of its modern reoccupation (Saville, 1907, 1910).

I speak of reoccupation, because much of Manabí, including the Jama River Valley, was largely depopulated in the early Colonial period. As described earlier, Spanish influence in coastal Ecuador dates from Pizarro's 1526 and 1531 expeditions from Panama, when the coasts of what are now Esmeraldas and Manabí provinces were explored (Zeidler, 1987). In the decades that followed, Puerto Viejo, established in 1535 in central Manabí, became the main entry point for expeditions to the sierra of Ecuador, as well as an important supply point for expeditions to Peru. Newsom (1995), in her study of the demographic impact of the conquest in Ecuador, estimates that by the end of the 16th century, 97.5% of the indigenous inhabitants of the Puerto Viejo region had perished to introduced diseases, conflict, and the forced labor practices of the Spanish. Only the Guayaquil region on the coast suffered a greater population loss, 99%. Newsom estimates that Esmeraldas, located further from the centers of conquest and commerce, lost some 71% of its native population. While there are no estimates for northern Manabí per se, a decline of the magnitude of Esmeraldas or greater can be assumed, and native culture soon disappeared from the region. These population decline figures are likely on the low side, given the difficulty of estimating the precontact populations, particularly those of the inland regions of the coast. Spanish population remained very light in this "backwater" of the Spanish colonial empire during the 16th century, and was largely oriented to the near coastal zone.

By the turn of the 20th century, some 100,000 inhabitants occupied all of Manabí, an area of 20,442 km^2. Saville lists the principal products of the province as cacao, coffee, sugar cane, tobacco, rubber, cotton, timber, textile plants, and vegetables. The primary product, however, was *tagua,* vegetable ivory from the *cadi* palm, which was exported to make buttons. Industries included agriculture (including cattle raising), fishing, and weaving, especially production of the world famous "Panama" hats from the native textile plant, *paja toquilla* (Figure 3.1).

Saville's expedition explored northern Manabí in 1908, traveling by horseback from Bahía de Caráquez to Cojimíes, at the frontier of Esmeraldas province. They rode along the beach, except in places where it was necessary to detour inland, around rugged coastal cliffs. One such detour was around Cape Pasado through the coastal forest to the town of Jama:

> [T]he trails were in a muddy condition, and traveling was difficult. Leaving the town [Canoa] we crossed the Río Muchacho at the base of the cape, and turned inland, taking a zigzag, winding trail through the dense forest, crossing and recrossing the Muchacho, the Tabuchilla, and Hachita rivers. . . . The streams, which we had to cross almost continually during the first part of our journey, were swollen; indeed, the greater part of the trail . . . was in the bed of creeks. In the part of the region in back of the cape, are a number of rich haciendas, for there is abundance of running water, and moisture rises from the heavy mists which blow in from the sea. Farther inland, about halfway to Jama, we ascend many slight, heavily forested mountain-ridges, which are undeveloped. Crossing these ridges, we come again into a country of haciendas . . . and we enter a good road, which continues until we reach Jama. (Saville 1910:22)

Saville was able to observe the coast near Cape Pasado on his return trip south, and notes that, in common with nearly the whole stretch of coast from Bahía de Caráquez to Cojimíes, "along nearly every mile of the coast where the beach-line is not rocky or precipitous, ancient remains can be found" (Saville 1910:22). This was certainly the case for the mouth of the Jama River, where quantities of potsherds were scattered along the coast.

The modern town of Jama was settled around the 1870s, on a broad expanse of fertile, alluvial land in the lower river valley (Figure 2.1). Not surprisingly, this prime location was also the site of prehistoric settlement; Saville noted that antiquities were often found in the steep right bank of the river, and the remains showed the presence of a large settlement. Closer to the coast were remains of a large salt-production area.

Saville did not travel into the interior of the Jama Valley, but continued northeast up the coast toward Coaque. Even today, travel from the town of Jama southeast to San Isidro, in the middle Jama River Valley, is not easy; the road is barely more than a dirt track, passable by truck or four-wheel drive vehicle during the dry season, but difficult even by horseback after the rains begin (Figure 3.2). The improved road to San Isidro angles southwest out of the valley, connecting San Isidro to the coast just north of Bahía de Caráquez.

Older residents of San Isidro describe how that town was settled in the 1920s, from the south, the Chone region. Like Jama, San Isidro and the other major towns of the middle river valley, Pechichal and Eloy Alfaro, were established on wide expanses of good alluvial lands, on the main channel or major tributaries of the Jama. And like Jama, middle valley towns sit atop archaeological sites (Zeidler and Pearsall, 1994b).

Figure 3.1 Bundles of harvested paja toquilla leaves

The Jama River flows northwest into the Pacific from headwaters in a range of coastal hills, passing through a zone of considerable ecological diversity (Figure 3.3) (Zeidler and Kennedy, 1994). The Jama is the largest drainage basin of northern Manabí province, measuring some 1612 km^2. In terms of broad climatic patterning, northern Manabí occupies a transition zone between the arid lands of the southern coast (central and southern Manabí and Guayas provinces, and south toward Peru), and the wetter northern coast (Esmeraldas Province, and north into Colombia). Further, rainfall increases from the coast inland. The lower reaches of the Jama River Valley (lowest 15–20 linear km) thus fall into the dry tropical climatic zone (average rainfall less than 1,000 mm/yr), with the driest areas to the southwest, while the interior portion of the valley is classified as semihumid tropical climate (1,000–2,000 mm/yr of rain), with the driest areas in the southwest interior. Rainfall is highly seasonal throughout the valley, falling predominantly from December or January into April, with some areas experiencing a period of *garúa* (drizzly rain) in October.

While topographic differences within the Jama River drainage are not dramatic, the interplay of rainfall and topography creates four different life zones (vegetation formations) (Zeidler and Kennedy, 1994). The low relief terrain of the lower valley is characterized by very dry tropical forest (coast southwest of Jama) and dry tropical forest (coast northeast of Jama), with dry forest extending inland into an area of rugged terrain called the Narrows. The higher elevations of the middle valley, inland of the Narrows, are characterized by dry premontane forest and humid premontane forest.

The landscapes of the lower valley include the mangrove estuary at the mouth of the river, the beach, the floodplains of the Jama River and its major tributaries, and the low coastal hills (100–200 m elevation). Floodplain soils are typically deep deposits of silt and silty clay loam favored for agriculture and settlement (modern

Figure 3.2 Scenes around the Jama River Valley: a. shrimp pools have considerably altered the landscape of the lower valley; b. mixtas, local transports consisting of wooden seats on a truck base, make the journey inland during the rainy season

and prehistoric), while the soils of the hills are more variable in composition and depth. Moving inland, the river flows through a deeply dissected gorge with no floodplain, the Jama Narrows, which is defined by a major NE–SW fault line. Hills rise from 300–600 m elevation; steep slopes are common. Inland of the Narrows, the valley opens up again. Most of this region (the middle valley from Pechichal to just

c. market day in San Isidro brings farmers in from the surrounding countryside; d. rolling topography characteristic of the middle valley around San Isidro

beyond Eloy Alfaro) falls into the humid premontane forest life zone; elevation varies between about 200 m and 600 m. As in the lower valley, large expanses of alluvium occur along the Jama River and its major tributaries, and were favored locations for prehistoric settlement. The uplands of the middle valley are a complex mosaic of soils.

Figure 3.3 Ecological and landscape diversity of the Jama River Valley: a. in the lower valley the Jama River flows through an area of drier forest; b. point where the river emerges from the Narrows, an area of steep terrain; c. lower hills and moister forest of the middle valley

Recent research by botanist David Neill and associates (1999) in the Mache-Chindul region—part of the coastal mountain range that extends southward to the Jama Mountains, as well as other research associated with the *Flora of Ecuador* project (e.g., Jørgensen and León-Yánez, 1999), is giving botanists new insights into phytogeography and vegetation of coastal Ecuador. Table 3.1 summarizes some of the tree species and genera that characterize the mature forests of the Jama region. Earlier work by Acosta-Solis (1961), Little and Dixon (1969), and botanist Robin Kennedy's and my collections during the Jama project (Pearsall, 1994a; Zeidler and Kennedy, 1994), have also contributed to our understanding of the life zones of the valley.

TABLE 3.1 TYPICAL TREE SPECIES (WITH BOTANICAL FAMILIES) OF MATURE FOREST IN THE JAMA RIVER REGION

Moister forests (tropical moist forest; semideciduous forest; humid premontane forest)

Mangifera indica (Anacardiaceae)
Aiphanes sp., *Attalea colenda, Phytelephas aequatorialis, Socratea* sp., *Bactris* sp. (Arecaceae)
Tabebuia chrysantha (Bignoniaceae)
Pseudobombax millei (Bombacaceae)
Swartzia littlei, Inga jaunechensis, Erythrina peippigiana, Centrolobium ochroxylum, Acacia pellacantha (legume trees)
Guarea cartaguenya (Meliaceae)
Castilla elastica, Brosimum alicastrum, Coussapoa villosa, Cercropia, Poulsenia armata, Ficus spp. (Moraceae)
Gallesia integrigolia (Phytolaccaceae)
Exothea paniculata (Sapindaceae)
Vitex gigantea (Verbenaceae)

Drier forest (deciduous forest, dry tropical forest)

Tabebuia chrysantha (Bignoniaceae)
Ceiba trichistandra, Eriotheca ruizii, Pseudobombax guayasense (Bombacaceae)
Erythroxylum glaucum (Erythroxylaceae)
Acacia spp., *Caesalpinia glabrata, Mimosa* spp., *Machaerium millei, Prosopis juliflora, Samanea saman, Pseudosamanea guachapele* (legume trees)
Ficus spp. (Moraceae)
Triplaris guayaquilesis (Polygonaceae)
Pradosia montana (Sapotaceae)
Vitex gigantea (Verbenaceae)

These life zones are not represented by intact primary forest today, however (Figure 3.4). Logging of hardwoods favored in construction and furniture making has removed most emergent giants and canopy tree species; cattle grazing has removed much of the understory of existing forest stands; local production of *carbón* (charcoal) is a continuous drain on remaining hardwood species; and extensive areas have been cleared permanently for agricultural plantations or improved pasture lands.

Using vegetation data collected during archaeological survey in the Narrows and the middle valley, Kennedy demonstrated that over 93% of land in the Jama Valley is utilized today for agriculture and grazing, with the remaining 7% mostly second-growth forest (Zeidler and Kennedy, 1994). Second-growth forests have high levels of representation of invader species such as wild papaya, *mojín (Triplaris guayaquilensis), laurel (Cordia alliodora), guasmo (Guazuma ulmifolia), balsa (Ochroma pyramidale),* and *sapán de paloma (Trema micrantha)*. This extensive 20th century alteration of the valley has implications for reconstructing past life-ways, as I will discuss in later chapters.

I discuss present-day agriculture in the valley in detail in Chapter 8. Coffee, cacao, plantains (cooking bananas), and bananas are important cash crops, and with cattle are mainstays of agriculture today. There is also growing interest in truck farming (tomatoes and peppers), commercial production of sweet corn under irrigation, and swine production. In the lower valley, fishing is still a viable, if small-scale industry, but the shrimp-growing facilities we saw under construction in the 1980s are now largely abandoned.

Figure 3.4 Habitat alteration in the region: a. much of the forest in the valley has been cleared and sown in pasture grasses; b. plantations of plantain (on donkey), coffee, and cacao (background) have largely replaced riverine vegetation; c. cattle husbandry is an important economic activity in the valley that has a profound impact on vegetation; d. the activities of huaqueros (looters) impact many archaeological sites but provide a source of income for valley residents

While the "boom" years of the 1970s are past, antiquities trafficking continues to be an economic enterprise in the valley. This involves both *huaqueros,* the men who "mine" sites, especially mortuary contexts, for whole vessels, figurines, and gold and silver ornaments, and the middlemen who sell pieces to national and foreign collectors. During the years of our fieldwork, archaeology also brought monetary benefits to the communities of the Jama River Valley in terms of jobs and other input into the local economy.

SNAPSHOTS FROM THE FIELD

During the decade that Jim Zeidler and I worked on and off in the Jama River Valley, we were privileged to work with great students, talented colleagues, and hardworking, generous local people (Figure 3.5). Our research would not have been successful without these collaborations. I will talk about the contributions of the ethnobotanical research team in the chapters that follow; here I would like to present a few "snapshots" of what it was like to live and work in the Jama River Valley during this project.

> This was a totally lost day. I wanted to go back up to the Narrows—to the house of Osvaldo Alciba, to collect maize data from a high maize field I'd seen earlier. I rented a horse and got a guide and went in, hoping the field had been harvested. It had, but Sr. Alciba wasn't home to show me how to get to the field. I went looking for the path

Figure 3.5 In the field during the Jama project: a. some of the 1982 ESPOL field crew, Jim Zeidler second from left, with Aurelio Iturralde and Cesar Veintimilla on his left; b. the author and Noéz Zambrano setting out on the land-use survey, 1982; c. University of Missouri paleoethnobotany students Andrea Hunter and Marcelle Umlauf learning to grind coffee, 1988; d. project botanist Robin Kennedy and helpers, 1990; e. a cheerful archaeology spouse, 1991; f. some of the 1991 field crew, Mary Jo Sutliff second from left, and Evan Engwall second from right

> with my 14-year-old guide, but never could get to it. So, I spent 1 hour on horseback looking for the house, 2 hours looking for the field, and 1 hour riding back. (Student's fieldnotes)

The logistics of conducting fieldwork in a rural area with poor roads, limited transport, and no phones are formidable. Even more daunting is moving different research teams (pedestrian survey, site testing, ethnobotany) in different directions in this setting. There were numerous days when things just did not work out: survey teams could not find the quadrat for the day and spent their time waiting to be picked up; the maize field visible from the river vanished up the hill; the horses promised in the afternoon were not available for the next day's plant-collecting trip. But balancing the "totally lost days" were days of discovery and insight, shot with the pleasure of hard work accomplished.

> Did I forget to mention the dust? (DP to MD)
>
> Mike carried the heavy loads and ate dust. (Note in margin of DP's fieldnotes)

There is a long tradition in archaeology of spouses accompanying researchers into the field. More than providing companionship and a sense of normalcy, archaeology spouses often make substantive contributions to the research program. My husband, Mike DeLoughery, spent part of two field seasons with me in the Jama River Valley, assisting with plant collecting and agricultural research. He drove me up four-wheel drive tracks to one remote farm after another, increasing the efficiency of the maize and yuca yield studies dramatically, and freeing me to focus on data collecting (rather than on whether I could turn the car on that rutted track).

> Today was another rainy day in stinky Pechichal. I swear there are at least five times as many pigs as people there, and they just run loose and poop everywhere. . . . We worked pretty late, and it was miserable. (Student's fieldnotes)

For academics and students more used to cracking books than toting bags of soil, the shear fatigue of fieldwork can come as a surprise. The hours are long, the work is physically demanding, and the pressure of too much to do in not enough time is constant. Add in ground slick with pig poop, and even the most enthusiastic ethnobotanist begins to wonder what she is doing here.

> Also passed a place where a bunch of chontilla palm fruits were hanging up by the house. Later, when we got to the Guzmán house, we were given a thick drink made from chontilla fruits and platanos (it was one of the more awful things I've had to politely drink!). (Student's fieldnotes)
>
> When I got back from dinner there was a big tarantula, several brown spiders, some giant cockroaches, and about 40 bats in my room. I went to sleep with bats circulating the air to keep me cool. (Student's fieldnotes)

Culture shock as a concept encountered in the classroom is one thing; living and working in its grip is something else. There is no way to know how a city-raised student or colleague will react to chickens in his bedroom, ankle deep dust on the road to dinner, or the command to jump from the truck, now, before it slides down the hill. When professors advise students to get some field experience before deciding on a career in archaeology, it is not just so they can decide if they like digging, survey, or flotation better.

> Went today on a trip to the Cevallos shrimp works, new fields, in the large bay at Cojimíes (Boca de Daule). This is good for the project, since this group of the Cevallos family (5 brothers and a sister) owns much of the land in the lower valley and Narrows. The main event was the inaugural ferry ride . . . many whiskies were drunk . . . a bottle of champagne was broken on the ferry . . . the trip back on the beach partway took 3 hours. Pretty rough road. (DP's fieldnotes)

Developing rapport is another one of those concepts that sounds easy, but turns out to have layers of complexity. Navigating the official bureaucracy of research permissions—to excavate, to export study collections, to collect plants, to export plant specimens—is straightforward, if time-consuming, but has nothing whatever to do with convincing a farmer to let you collect on his land. Jim Zeidler broke the ground for the project, putting in countless hours explaining us, chatting up landowners, doing favors, and becoming part of the San Isidro and Jama communities.

> We [Noéz Zambrano and DP] did a long walking tour today (15 km) into the higher elevations around the Estero Cañaveral and E. Congrejo valleys, getting a good look at remnant montaña vegetation, higher elevation fields, pasture land, and the headwater areas of the two esteros. . . . Noéz explained that as one gets higher, towards the tops of the cerros, humidity begins to increase. . . . [D]uring the ascending part of the walk we talked about agriculture; Noéz told me how different races of maize will cross if planted too close. . . . [F]ields, abandoned because of the death of the crops or lack of care by the owners, can be cleared and reseeded after 3 years. . . . [W]hile descending from Buenos Aires to the Las Brisas road, I asked Noéz about the native trees in the zone. He knew some, and identified others as we encountered them. (DP's fieldnotes)

This excerpt is from one of the most enjoyable days I spent in the Jama River Valley, October 12, 1982. The notes I wrote afterward fill six single-spaced pages, but they do not capture what the day was really like. The cool, damp air of the ridgetops; the hot sun and dust of the dirt tracks; bamboo and thatch houses clinging to the hillsides; coffee and cacao sheltered under broad-leafed trees; cattle grazing among palms; curious children dogging our steps. I had been in San Isidro less than a week. Noéz was my guide all that first field season; he explained me to farmers, surprised me with oranges when we were hours out of town, swam our horses across a swollen Jama River, and showed me where we actually were on the aerial photographs. He and his family, like many local people, made us feel welcome and helped to make the project a success.

4/Recovering the Archaeobotanical Data

SITES TESTED DURING THE PROJECT: CULTURAL CHRONOLOGY AND METHODS

Our knowledge of plant-people interrelationships in the Jama River Valley comes largely from plant remains recovered from archaeological sites tested during the project, and from more extensive excavations at San Isidro, the middle valley ceremonial center. In all, archaeobotanical remains were recovered from 14 sites (Figure 4.1). In addition, vegetation change in the San Isidro area was investigated by sampling a stratigraphic cut in the Jama River alluvium—the Río Grande profile—for phytoliths, and by coring in small ponds for pollen.

The project study region, shown in Figure 4.1, is the central axis of the Jama River Valley, comprising some 785 square km. As discussed in the last chapter, the life zone ecology of the region varies from dry tropical forest along the Pacific coast, to humid premontane forest at higher elevations in the interior of the valley. While the transition from the drier coastal zone to the more humid interior is gradual, when topography and soils are taken into account, three distinctive physiographic zones can be defined (Zeidler and Kennedy, 1994). These zones were used by Zeidler to stratify the valley into three strata for the archaeological survey (Zeidler, 1995). I used the same stratification for the agricultural study, described in Chapter 5. Stratum I is the semiarid coastal plain, Stratum II the subhumid coastal cordillera, or Narrows, and Stratum III the humid upland valley. For the survey, Stratum I was further divided into substrata comprised of three large pockets of alluvial bottomland, two shoreline areas, and the extensive nonalluvial uplands; and Stratum III was divided into 14 pockets of alluvial land and the nonalluvial uplands. Stratum II was comprised only of nonalluvial lands. Test excavations were placed in selected sites in each stratum.

Archaeobotanical data are only as good as our understanding of the cultural context and age of the samples. The age of archaeobotanical samples analyzed during the Jama project was determined, ultimately, by radiocarbon dating. No food remains were directly dated, however. Radiocarbon dating established the age of three wind-deposited volcanic ash falls, or tephras, that blanketed the valley in pre-

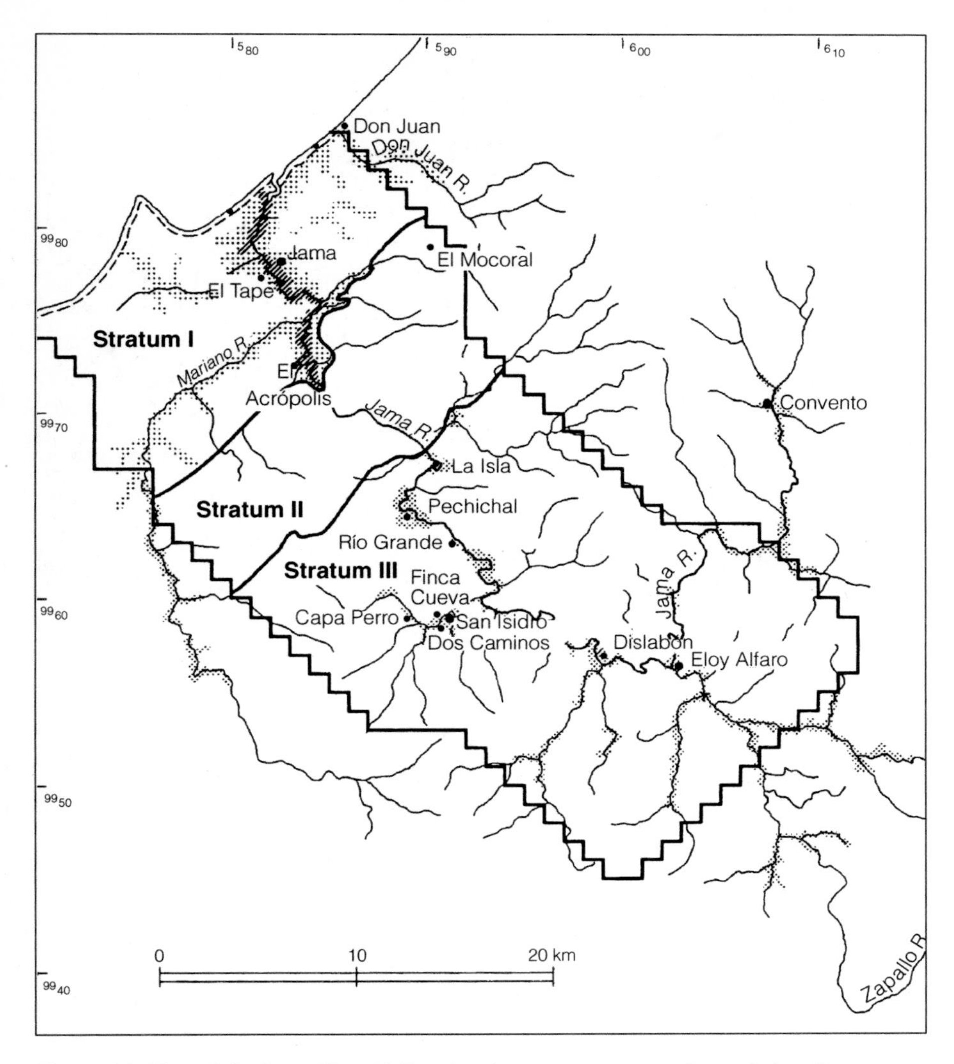

Figure 4.1 Map of the Jama River Valley showing survey strata and tested sites. Base map adapted from Figure 2.7 in Regional Archaeology in Northern Manabí, Ecuador, *ed. Zeidler and Pearsall. Used with permission.*

history and were visible in many test-pit profiles and in the Río Grande profile (Figure 4.2). Dates on charcoal also established the age of the debris left by the first settlers of the valley, people producing early Piquigua ceramics, and allowed Zeidler to date the refined chronology of the 2,000-year-long Jama-Coaque sequence (Muchique 1–5) (Zeidler et al., 1998). In all, 37 radiocarbon determinations were analyzed to produce the 3,600-year cultural chronology displayed in Table 4.1.

The age of archaeobotanical data from a deposit is thus established by knowing the kind of pottery recovered in the deposit. This establishes the cultural affiliation of the plant remains. Jim Zeidler and Mary Jo Sutliff (1994) determined the cultural affiliations of deposits. In this book, I will refer to the botanical data using cultural affiliation, rather than chronometric age.

Figure 4.2 Jama River cutbank, showing the tephra deposits (white horizontal strata)

Sites were tested by Zeidler and the archaeology field crew (Zeidler, 1994a, b). In most cases, test excavations had two goals: to recover ceramics and other materials necessary to establish the cultural chronology, and to recover a systematic sample of floral and faunal remains to reconstruct past foodways. Long, continuous stratigraphic sequences are essential for chronology building, and so many of the test excavations are of this type. For example, excavations in Sector XII/Area C at San Isidro revealed a 5.5 m deep sequence with 32 distinctive deposits that eventually provided the "master" ceramic sequence for the region. Excavations were also placed at San Isidro to investigate the sequence of construction of the large central mound.

Excavation units typically measured 2 × 2 m. The first unit in a site was often excavated in arbitrary levels (usually 20 cm). The profiles exposed in the unit revealed the natural archaeological deposits. Sediment samples were taken from the stratigraphic units of these profiles for grain size analysis, pollen analysis, and phytolith analysis. If time permitted, adjacent units were then excavated by the natural stratigraphy revealed in the profiles. The field teams also took advantage of looters' pits, river cutbanks, and other exposures of a site's natural stratigraphy by cleaning the exposed profiles and placing excavation units adjacent to them. This permitted excavation by natural stratigraphy. In some cases, the "cleaning" of exposed profiles required removal by natural deposit of large amounts of sediment, which were subsequently floated to recover charred botanical remains.

One constant volume soil sample was taken for flotation from each natural or arbitrary excavation level, with a separate sample taken from each feature exposed during excavation (Pearsall, 1994a). Large features, like the bell-shaped pit excavated at the Pechichal site, were sampled by depositional episodes. This strategy, referred to as blanket sampling, makes the process of taking samples routine: the field team does not have to evaluate whether the deposits are rich in plant or animal remains during excavation (Pearsall, 2000). We started with 20-liter samples, but

TABLE 4.1 CULTURAL CHRONOLOGY[1]

Cultural Component	Ceramic Phase	Disjuncture	Modal Age Range cal. B.C./A.D.
Campace?	Muchique 5		A.D. 1430–1640
		Spanish Conquest	
Jama-Coaque II	Muchique 4		A.D. 1290–1430
Jama-Coaque II	Muchique 3		A.D. 880–1260
Jama-Coaque II	Muchique 2		A.D. 420–790
		Tephra III	immediately prior to A.D. 420
Jama-Coaque II	Muchique 1		240 B.C.–A.D. 90[2]
		hiatus	lasts 250–500 years
		Tephra II	750 B.C.
Chorrera	Tabuchila		1300–750 B.C.
		hiatus	lasts ca. 500 years
Valdivia 8	Late Piquigua		ends 1880 B.C.
		Tephra I	
Valdivia 8	Early Piquigua		begins 2030 B.C.

[1]Reproduced by permission of the Society for American Archaeology from *Latin American Antiquity* 9(2), 1998.
[2]Zeidler suggests that Muchique 1 may begin close to 350 B.C. and continue until the Tephra III event.

increased sample size in the later years of the project in an attempt to increase the amount of material recovered. I describe flotation procedures below. Based on their analysis of the ceramics, many samples from arbitrary excavation units were eventually assigned to cultural phase by Zeidler and Sutliff (1994).

Table 4.2 summarizes the archaeobotanical database by stratum, site, and cultural phases. The five sites that contain deposits affiliated to early Jama-Coaque II, the Muchique 2 phase—El Tape, Pechichal, San Isidro, Finca Cueva, and Dislabón—will be discussed in more detail in Chapter 6. Here I provide brief descriptions of the locations and characteristics of all tested sites for which there are botanical remains.

Archaeobotanical samples were recovered from four sites tested in Stratum I, the coastal plain: Don Juan (M3B2-001), El Tape (M3B3-002), Jama (M3B3-003), and El Acrópolis (M3B3-012) (Zeidler, 1994a). Don Juan is a multicomponent site located in the lower reaches of the Río Don Juan, just behind a small estuary at the river's mouth. While not strictly part of the Jama River drainage, the alluvium in which the site is located is connected to the alluvial deposits at the mouth of the Jama by a thin strip of alluvium. Much of Don Juan has been destroyed by modern habitation, road building, and construction of shrimp ponds. El Tape is a multicomponent site located on the west side of the Jama River opposite the modern town of Jama. It is a deeply stratified site on an old alluvial terrace and is being destroyed by river erosion and looting activities. The Jama site is located on the east side of the Jama River beneath the modern town. It is directly across the river from El Tape, and like that site occupies an extensive stretch of old river alluvium. El Acrópolis is a mound site, dating to Late Muchique 3 and Muchique 4, located in the eastern part of Stratum I. It is in an old alluvial terrace on the west side of the Jama River, but not on the river itself.

Only one site has been tested in Stratum II, the Narrows: El Mocoral (M3B4-031). El Mocoral is a multicomponent, nonalluvial site located during the 1990

TABLE 4.2 LIST OF SITES WITH ARCHAEOBOTANICAL REMAINS

	Site Number	Site Name	Term. Valdivia Piquigua	Chorrera Tabuchila	Jama-Coaque I Much 1	J-C II (Undivided)	J-C II Much 2	J-C II Much 2/3	J-C II Much 3	J-C II Much 4
Stratum I										
	M3B2-001	Don Juan		x	x				x	
	M3B3-002	El Tape	x	x	x	x	x			
	M3B3-003	Jama						x		
	M3B3-012	El Acrópolis							x	x
Stratum II										
	M3B4-031	El Mocoral		x	x	x				
Stratum III										
	M3B4-007	La Isla						x	x	
	M3B4-011	Pechichal					x			
	M3D2-001	San Isidro	x	x	x	x	x		x	
	M3D2-006	Río Grande		x						
	M3D2-008	Dos Caminos		x						
	M3D2-009	Finca Cueva			x		x		x	
	M3D2-056	Dislabón			x		x			
	M3D2-065	Capa Perro	x						x	
	M3D2-091	Eloy Alfaro	x	x						

survey. The site is oval in outline, some 300 m × 100 m in size, and lies at 250 m elevation, on sloping land above the Estero Mocorón, which flows northwest into the Río Don Juan drainage.

Botanical remains were recovered from nine sites tested in Stratum III, the middle valley zone: La Isla (M3B4-007), Pechichal (M3B4-011), San Isidro (M3D2-001), Río Grande (M3D2-006), Dos Caminos (M3D2-008), Finca Cueva (M3D2-009), Dislabón (M3D2-056), Capa Perro (M3D2-065), and Eloy Alfaro (M3D2-091) (Zeidler, 1994b). All are located on the large, discontinuous expanses of old alluvium that characterize the Jama River and its major tributaries in Stratum III.

The first expanse of old alluvium in the middle valley, located on the edge of the Narrows, contains several sites. La Isla, located in a loop of the Jama River, was tested. There are a series of U-shaped mounds at this site. The Pechichal site is located upstream on the western side of the Jama River, on an old alluvial terrace beneath the modern town of that name. Five large storage pits were eroding out of a cutbank; one was excavated and provided the richest biological record for the valley. The Río Grande site is located on the next expanse of Jama River alluvium, in the Valle Alegre area, about 5 km northeast of San Isidro. The site is at the northern end of the Valle Alegre alluvium, on the western side of the river. The Río Grande geological profile is located adjacent to the Río Grande site.

San Isidro is the mound site that is the central place of the middle valley. Dos Caminos, Finca Cueva, and Capa Perro are all located on the alluvium of the Río Cangrejo, a major tributary of the Jama (or in the case of Capa Perro, on a tributary of the Río Cangrejo). These are all deep, multiple component sites. Dos Caminos and Finca Cueva are located across the Río Cangrejo from each other and just upstream from San Isidro, in an area where the alluvium narrows somewhat. Capa Perro is located about 2 km upstream of San Isidro, in an area of still narrower alluvium. The broadest expanse of old alluvial terrace is covered by the modern town of San Isidro, and the site of that name.

The Dislabón site is a multicomponent site located on the eastern side of the Jama River in a large pocket of alluvium upstream from the Río Cangrejo/Río Jama confluence. The site is situated in the broadest part of the alluvium, opposite the confluence of the Río Cucuy and the Jama River. The final site tested in Stratum III, Eloy Alfaro, is located in the Jama River alluvium at the modern town of that name.

RECOVERING MACROREMAINS: FLOTATION AND IN-SITU SAMPLES

Water flotation of large soil samples taken from test units at the sites described above was employed to recover a systematic sample of charred plant remains (macroremains) and faunal materials. While there are advantages to recovering plant materials during excavation (one has more precise information on the depositional context of the materials), in reality only large plant remains are visible to the naked eye during excavation. If paleoethnobotanists relied only on in-situ finds for reconstructing human-plant interactions, we would know little about, for example, the use of small seeds or food remains that occur in highly fragmented form (Pearsall, 2000). Small animal bones are also routinely missed during excavation, and not caught in 1/4" excavation screens. To recover a picture of past plant use unbiased by size requires either fine sieving of sediments or water flotation (Figure 4.3). Only

J. Zeidler

Figure 4.3 Flotation crew in action

those plant remains that became charred due to cooking accident, burning of trash, or because they served as fuel are preserved in sediments that are seasonally wetted and dried, such as those found in the Jama River Valley.

Flotation, or "washing the dirt" as one crew member called it, was the method we employed for systematic recovery of archaeobotanical materials. Sediments were often very clayey, making fine sieving impossible; indeed, clays took a long time to disperse, and flotation was a tedious business. That we have good samples from many sites is due to the hard work put in by a succession of two-person flotation crews made up of local women and men from the towns of San Isidro and Jama. Prior to the 1988 field season, I trained ESPOL student Cesar Veintimilla in IDOT-style flotation, and he supervised the flotation crew that season; University of Missouri graduate students Marcelle Umlauf and Eric Hollinger did this duty during the summers of 1989 and 1990, respectively; and lab director Mary Jo Sutliff took on the duty for the final 1991 field season. University of Illinois graduate student Evan Engwall supervised the flotation of soil from sites he tested during 1994. The first excavations at the San Isidro site (1982, 1983) were carried out as ESPOL field schools; Veintimilla processed those flotation samples using ESPOL's SMAP-style flotation device.

The IDOT flotation system was developed by paleoethnobotanist Gail Wagner in the 1970s during a project for the Illinois Department of Transportation (hence the name IDOT). The equipment for this system and the principle behind its operation are simple. Fine-gauge wire cloth is wrapped around a wooden frame, as shown in Figure 4.4. If sediments are fine-grained, 0.5-mm mesh is used. This mesh is large enough to allow silt and clay soil particles to pass through, but small enough to catch the smallest seeds typically encountered. The wooden frame

Figure 4.4 Example of an IDOT-style flotation device

allows for easy construction using hand tools, and provides some buoyancy for the device during flotation. The second piece of equipment, the hand sieve, is made of 0.25-mm wire cloth, sometimes referred to as carburetor mesh, soldered onto a circle of wire. The ends of the wire are inserted into a wooden handle.

The IDOT flotation device is a type of manual flotation system (Pearsall, 2000). The procedure followed during the Jama project, briefly, was as follows. The crew of two carried the device and a bag of soil into shallow water. One crew member immersed the flotation device partway in the water, while the other poured a few liters of soil into it. The first crew member shook the device in the water—either up and down, or side to side—to break up the soil, allowing plant remains to float to the surface. The second crewmember scooped the floating material out with the hand sieve, and tapped it onto a cloth square positioned on a stationary platform. This procedure was repeated until the entire sample was processed. The device was emptied of nonbuoyant material periodically to reduce the weight and increase the efficiency of the soil "washing."

In principle, most botanical materials are either light enough to float in water (small seeds) or contain air pockets that provide buoyancy (most wood charcoal, small root remains). Most of the material floats free and is scooped out, forming the *light fraction*. Palm nutshell and some kinds of wood are very dense, however, and do not float to the surface. These materials can be scooped off the bottom by raising the device out of the water, dropping it quickly back in, and scooping out the materials, which are raised off the bottom by the action of the water. (If flotation is carried out in an elevated tank, rather than a river as in the Jama project, an aquarium siphon can be used to siphon semibuoyant materials off the bottom of the device.) In practice, it was difficult to get all dense materials off the bottom of the device during flotation; charred materials in the nonbuoyant, or *heavy fraction,* were more often removed by hand in the field laboratory, after samples were passed through a 1.0-mm geological sieve. Bones and other faunal remains were captured in the heavy fraction and removed for analysis as well.

When analysis of 1982, 1983, and 1988 flotation samples revealed few charred materials per liter of floated soil (i.e., low density of remains), I recommended that we increase sample volume from 20 liters of soil to 30. This higher volume was used during subsequent field seasons. The larger samples placed a heavy burden on flotation crews, however: the 50%+ larger samples typically took an hour and a half or more to process. Although just part of the sample was poured into the device at a time, the extended "washing" needed to disperse clays allowed botanical materials to waterlog, with many ending up in the heavy fraction. The large samples were also difficult to dry, and were sometimes floated moister than ideal, again resulting in decreased buoyancy. While reduced buoyancy was not a problem for larger materials that could be easily sorted from the pebbles and bones in the heavy fractions, small seeds were very difficult to pick out. Recovery of small archaeological seeds may have been affected by these problems. We measured the recovery rate for small seeds by adding a known quantity of "exotic" charred poppy seeds to a sample prior to flotation, and then counting the number recovered. Recovery ranged from 0% to 84%, with an average recovery rate of 38% for 57 tests.

In hindsight, perhaps I should have recommended a switch to a machine-assisted flotation system after the 1988 season. SMAP-style flotation devices, in which water pumped through the sample provides lift for the botanical remains, are well-suited to clayey soils and large sample volumes (Pearsall, 2000). The logistics of building our own SMAP system and transporting and storing it in San Isidro, however, seemed formidable at the time. Added to this was the problem of transporting soil from sites to a stationary flotation setup: the IDOT system was easy to transport to the vicinity of the site being tested. Whatever the best decision would have been, the float crews processed over 250 samples using the IDOT device, and produced the data that form one of the foundations of our understanding of plant and animal use in the Jama River Valley.

POLLEN AND PHYTOLITH SAMPLES

Small sediment samples for pollen and phytolith analysis were taken from test pit profiles and other exposures of long stratigraphic sequences during the project (Pearsall, 1994c). After a fresh surface was cut in a profile, two sediment samples of 100–200 g each were taken from each deposit, beginning at the base of the profile and working upward (Figure 4.5). Working upward minimizes contamination of deposits with sediments dislodged from already sampled strata. Sampling was discontinuous; in other words, samples were taken from the center of each deposit and typically were not contiguous with the next sample. Care was taken in sampling thin deposits not to cross the deposit boundaries while sampling. Sterile, unopened plastic bags were used to hold sediments, and tools (a trowel or the sampling device shown in Figure 4.5) were cleaned between samples to minimize contamination.

In addition to sampling profiles of tested sites, the Río Grande profile, a naturally deposited sedimentary sequence exposed along the Jama River ca. 3.5 km

J. Zeidler

Figure 4.5 Sampling a profile for phytoliths and pollen

northeast of San Isidro, was also sampled this way by project geologist Jack Donahue of the University of Pittsburgh (Donahue and Harbert, 1994).

I discuss the nature and interpretation of microfossil evidence in more detail later, but a word here about the differences between pollen and phytoliths is in order. Phytoliths are a type of inorganic deposit in plant tissues (Pearsall, 2000). Silica that forms phytoliths is carried up from groundwater as monosilicic acid and deposited in cells—perhaps as a waste product, perhaps to serve as support or to perform an antiherbivary function. Distinctively shaped bodies are formed in many plants. When the plant decays or is burned, phytoliths become deposited in sediments. They do not decay, being inorganic, but may dissolve in sediments of very high pH. Pollen grains are similar in size to many phytoliths, but serve a very different function: they are the male gametophyte, essential in reproduction in the higher plants. The living portion of the pollen grain is protected by layers of tough natural organic material. Even pollen grains can be destroyed, however, especially if their outer surface (the exine) is eroded by contact with soil particles or is broken by successive wetting and drying cycles of sediments. Bacteria and fungi also attack pollen grains. Decay is retarded in anaerobic conditions, especially in waterlogged sediments that are acidic (low pH).

On the advice of palynologist Vaugh Bryant, Jr., whose graduate student Laurie Zimmerman analyzed the Jama pollen samples, we dried the sediment samples for pollen analysis in the field, using a low-temperature drier constructed for drying botanical voucher specimens. Even if bacteria and fungi are not active in sediments in situ, the process of sampling exposes sediments to light and air, stimulating the agents of decay. Drying is one way to reduce the activity of these biological agents. Phytolith samples were not dried until just before laboratory processing, since phytoliths do not decay.

I bring up these differences in pollen and phytolith preservation and field treatment of samples because Zimmerman's analysis of 53 pollen samples taken from four sites tested in 1989 and earlier (San Isidro and three sites in Stratum I) revealed very poor pollen preservation (Zimmerman, 1994). Few pollen grains were present, and none provided insight into past foodways or environment. In all likelihood, this situation was not the result of postsampling loss of pollen, but caused by conditions within the in-situ deposits. Zimmerman's assessment—that pollen loss resulted from the floodplain context of the four sites and the marked wet and dry seasons of the valley—led us to de-emphasize pollen analysis of archaeological sediments in subsequent field seasons. All tested sites, with the exception of El Mocoral (M3B4-031) in Stratum II, are located in floodplain settings; there was little reason to expect better pollen preservation at other sites.

An alternative strategy for recovering pollen was tried during the 1990 field season. Zimmerman extracted sediment cores from two small ponds and a swampy area in the vicinity of San Isidro. If ponds or swamps do not dry out seasonally, and pH values of bottom sediments are in the acidic range, pollen contained within sediments is usually in excellent condition. Pollen becomes deposited on the landscape, including the surfaces of ponds and swamps, primarily through the action of wind. While plants pollinated by animal vectors—bees, birds, and the like—are usually underrepresented in sediments, wind-pollinated plants will be well represented, and provide valuable insights into the nature of vegetation in the region surrounding the pond or swamp.

While lake coring has proven to be a valuable tool in understanding the relationships of people and landscapes in the lowland tropics (Pearsall, 1995a), in the case of the Jama project, the approach proved of limited use. Zimmerman was only able to extract short cores from the swamp and ponds because clayey sediments were encountered that the coring device would not penetrate. The cores did not sample ancient sediments, and so the records were not useful for the project goals.

5/Modern Flora and Agricultural Studies

Goals and Methods

As I discussed in Chapter 2, one of the ultimate goals of the Jama project is to investigate the evolution of prehistoric agricultural systems in the dry tropical forest of coastal Ecuador. In the preceding chapter, I outlined the methods employed to recover botanical data essential for reconstructing past plant-people relationships in the valley, and for testing models of agricultural evolution, such as that proposed by Rindos (1984). In this chapter, I describe field methods used to accomplish two other important steps in the research process: (1) documenting the "natural" environment of the study region and establishing analogs between different vegetation formations and microfossil indicators, and (2) modeling agricultural productivity in the valley. These steps are essential, respectively, for documenting the impact of people on the environment of the Jama Valley, and evaluating the extent to which differences in agricultural productivity influenced prehistoric settlement.

STUDY OF THE MODERN FLORA

Documenting the nature of the present-day flora and evaluating the extent to which it reflects the flora of the valley during prehistory accomplishes several goals. Traditional agriculturalists live in intimate contact with the natural world, and we cannot hope to understand how people lived in the Jama Valley for 3,600 years without understanding the workings of this world, and the resources and challenges it presented. This world does not exist intact today. However, clues to what the valley looked like to its first settlers do exist in remnant forest stands. How the dry tropical forest responded to the pressures of agriculture can be seen in patterns of regrowth of abandoned fields. Differences in vegetation due to elevation, aspect, slope, and groundwater resources can still be seen in hedgerows, trees left as crop cover, and modern uses of the land.

Identifying archaeobotanical remains depends on having comparative plant materials: wood, fruit, seed, and root specimens for macroremain identification, fruit and leaf specimens for phytolith analysis, and flowers for pollen extraction (Pearsall, 2000). Plant collecting is not the only way to get these specimens, but it is how you know *what specimens you will likely need.* Understanding how an assemblage of phytoliths from a sediment sample reflects past vegetation depends on establishing analogs: soil samples matched to stands of known composition. One has to understand the nature of the vegetation to set up useful analogies.

How can these distinctive, yet interlocking goals of the study of modern flora be accomplished? Can an archaeologist do this work? The answer to the first question is, through fieldwork planned around nature's rhythms; and yes, an archaeologist—trained as an ethnobotanist—can do this work, but only if she or he can commit the time. In the case of the Jama project, MU botanist Robin Kennedy took up the study of modern flora beginning in the 1990 field season, after I realized I could not spend enough time in the field to finish both the agricultural and floristic studies (Pearsall, 1994a, b; Zeidler and Kennedy, 1994).

How one studies the modern flora depends on the scale of the study, the research questions, and prior knowledge of the flora. In the case of the Jama Valley, there is no published flora for Manabí, only broad life-zone studies (i.e., Acosta-Solis, 1961) and a work on arboreal species of Esmeraldas, the province to the north (Little and Dixon, 1969). We therefore took a basically descriptive approach. We sought to identify the plants that dominated the flora—the trees of the dry coastal zone, the subhumid coastal cordillera, and the semihumid and humid uplands—and to describe the distinctive vegetation formations of the valley—that is, to document which species occurred together.

These goals were accomplished during six field seasons, spanning a nine-year period. Observations and collections were made in every month of the year except February, May, and September; this gave coverage during both the rainy (January–April) and dry (May–December) seasons. Fieldwork was organized this way because trees of the dry tropical forest flower following different seasonal patterns, and flowering specimens are essential for precise scientific identifications of plants.

Plant collecting and observation are a lot like archaeological survey—a kind of planned wandering over the landscape. While the pedestrian survey crews walked randomly selected quadrats, the plant-collecting teams relied on the knowledge of local people to locate fragments of intact forest. Most of the forest remnants to which our guides took us cannot be called "primary forest," since selective logging and firewood cutting is practiced, and cattle grazing has altered the understories. However, these "old-growth" forest remnants provided comparative specimens of many taxa that would have been more abundant in the past. The ethnobotanical knowledge of our guides, and that of older residents to whom we showed our specimens, was essential in providing us with the common names of trees (including those that had disappeared due to logging), their uses, and habitat preferences. Common names proved especially useful in helping Kennedy identify specimens that were never encountered in flower (Zeidler and Kennedy, 1994). Collections were also made in second-growth forest (regrowth of formerly cleared areas). Observations were made in each stratum, but we collected most of our specimens in Stratum I and Stratum III.

Methods varied somewhat from season to season, but were basically as follows. Material to make three or more voucher, or herbarium, specimens was collected for each species. Herbaria are institutions that store dried plant specimens for scientific study. In Ecuador, foreign researchers are required to deposit specimens at a herbarium in the country as part of receiving permission to collect; our specimens were deposited at the Herbario Nacional del Ecuador (QCNE) in Quito. (Requirements vary by country, so it is important to find out what the regulations are before beginning research.)

For arboreal species, the ideal voucher specimen included several leaves, showing their arrangement on the branch, and a number of flowers, essential for identification of the specimen. If present, fruits were included. Extra fruits were collected for the macroremain and phytolith comparative collections, and a wood specimen was cut. While wood from the trunk is preferred for comparative specimens, we cut 2" diameter branches for this purpose.

Specimens were sometimes pressed in the field, but we also accumulated material tagged with a collection number in a sack, and pressed specimens back at the field laboratory (Figure 5.1). The interested reader can consult Alexiades (1996), Martin (1995), and Pearsall (2000) for detailed instructions on how to press plants. Basically, one arranges the material in a newspaper, labeled with the collection number, in such a way that the features of the specimen are displayed clearly. In other words, leaves and flowers are laid out flat, with not too much overlapping of parts. Extra flowers and fruits are always useful, and can be placed in a folded paper or envelope in the pressing paper. The ethnobotanical comparative specimens—the fruits, leaves, and wood specimens that would be used to identify archaeobotanical remains—are placed in separate envelopes bearing the same collection number as the voucher specimens.

We then dried our specimens using either a field drying rack or a succession of blotters to draw out the moisture in the specimens. When specimens were dry, they were bundled securely in sets and placed in large plastic bags, which were closed tightly to keep specimens from rehydrating.

Herbarium specimens, like archaeological collections, are useless without provenience information. Before our collections were deposited, tags were placed in each specimen paper with full provenience information: project name, description of collection locality, map coordinates, associated plants, and collector name and collection number.

Collecting plants was an important aspect of the study of the modern flora of the Jama Valley, but not the only one. Equally important were observations on plant associations: what species grow together, and in what abundances. No primary forest exists in the Jama Valley, so multiple observations of associations in old-growth forests form the basis of our understanding of "natural" stand composition. Equally important were observations of secondary growth and habitually disturbed vegetation; these provide our analogs for extensively human-altered habitats in the valley.

Collecting this information begins by listing the species one observes—often by common name or collection number, since identifications may be made later—in a defined area. The next step is to estimate the abundance of the various species. This may be very informal, simply noting which species are the most abundant, or involve counting the trees of a certain diameter or estimating their canopy cover. To estimate canopy cover, one rotates in a circle, with one arm raised about 40 degrees above horizontal, and estimates the percentage (of a circle) of sky at that height that is covered by each species. Then one notes the composition of the understory and groundcover. Table 5.1 shows such observations for two areas in the Campamento region of the lower valley.

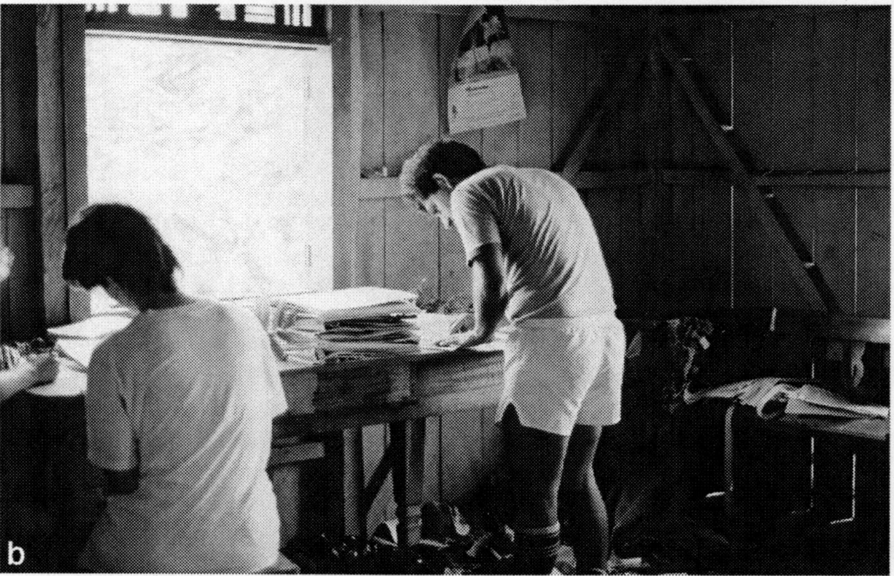

Figure 5.1 Collecting and processing plant specimens: a. field pressing; b. preparing voucher specimens in the lab

The final step is taking a surface soil sample. I usually pace off a 10 × 10 m square in a "typical" part of the stand being studied. Using that most useful tool, the archaeologist's trowel, I walk around the square, scrape the leaf litter from a spot, and put a small scoop of soil in a plastic bag (Figure 5.2). After 20+ repetitions, this results in a 100 to 200 g composite sample of the site. The sample is mixed by shaking the bag periodically. It is labeled with the date and provenience information. If

TABLE 5.1 EXAMPLE OF ESTIMATING ARBOREAL STAND COMPOSITION[1]

Campamento area, due west of Jama. Plant collecting in the Quebrada El Páramo, about 1 mile from the Campamento crossroads.

Area 1: south facing slope, dry forest, above an area planted with platano and yuca. Fairly closed canopy, but with few large trees; abundant herb cover; some areas with dense shrub and lower regrowth.

> General observations: *palma de Africa*, several; Q,[2] very common, dominant; L, a couple; S, two; V, one; N, many
> Upper story canopy: Q, 90%; L, 2%; unknown 3%
> New plants observed in the area: 1669,[3] 1664, 1661, 1657

Area 2: north facing slope, just down from a platano and yuca field; lower part of the slope.

> General observations: lots of vines; F, one mature tree, now a stump; Q, predominant upper canopy cover; L, medium abundance in lower canopy; *negrito*, a couple in the lower canopy; *bruja*, 1 large tree in lower canopy; wild papaya, 1 in lower canopy, 1 in upper; N, medium abundance in lower canopy; 1660, 1 huge tree in upper canopy; K, several, medium abundance in lower canopy; 1665, sparse in lower canopy; 1668, present; unknown Bombacaceae
> Upper story canopy: 1660, 3%; unk Bombacaceae, 3%; wild papaya, 2%; Q, 45%. Estimated; view obstructed by dense medium canopy growth
> New plants observed in the area: 1668, 1665, 1658

[1]From Pearsall field notes, pp. 329–331, 1988 field season.
[2]The letters refer to entries in a sketchbook of plants collected on previous trips.
[3]The numbers refer to plant voucher specimens.

the sample will be used for both pollen and phytolith analysis, it is dried as described in Chapter 4.

THE AGRICULTURAL AND LAND-USE STUDIES

In order to understand how the diverse topography and soils of the valley affect the agricultural potential on a localized level, I studied yields of maize and yuca under traditional cultivation practices (Figure 5.3). Maize is documented prehistorically on the coast, and we anticipated recovering it in the tested sites (and did so). Yuca is grown today in the valley, and was probably grown widely on the Pacific coast in prehistory. The varieties grown today are all "sweet" (low toxicity) yuca, as is typical for areas west of the Andes. A variety of other root/tuber foods could have been grown (and we identified several in the tested sites), but yuca was the only indigenous root/tuber food under widespread cultivation in the valley today. Maize and yuca productivity was studied during six field seasons of the Jama project by myself and several student research assistants (see Chapter 2).

Several hypotheses structured the way we collected the yield data and my subsequent analysis of it. Crop productivity is dependent on a number of factors, chief among them soil fertility, and the abundance and timing of rainfall. Little land in the Jama River Valley is devoted today to *ciclo corto,* the annual rainfall cropping of maize, yuca, and various vegetables. Most land is in pasture or used for cultivation of cacao, coffee, plantain, and banana (CCP for short). Maize and yuca both play a role in the local economy and food habits, however, and are grown in scattered plots

Figure 5.2 Taking surface soil samples

around the valley, in deep, periodically renewed, alluvial soils as well as on thinner upland soils of differing fertility. The first hypothesis, then, is that alluvial lands are more productive for maize and yuca agriculture than nonalluvial lands, and that the more fertile upland soils will outproduce the less fertile ones.

There is great diversity in the slope, aspect, and elevation of nonalluvial lands used for *ciclo corto* plantings, especially in Stratum II and Stratum III, where plantings range from 200 to 600 m. In traveling around the valley, especially in the area to the southwest of San Isidro, it became apparent to me that there was more moisture at higher elevation in some areas. CCP appears on ridge tops and the upper slopes of hills, for example, but less commonly on the lower slopes, except where such lands border on the river. Maize plantings are often associated with nonalluvial CCP plantings, or remnants of uncleared forest at higher elevations. A second hypothesis is that maize productivity in the uplands (nonalluvial soils) increases as elevation increases.

Finally, as discussed earlier, Zeidler divided the valley into three strata for purposes of archaeological survey: Stratum I, the drier lower valley, Stratum II, the Narrows, the area in which the Jama River cuts through rugged terrain with no alluvial soil deposition, and Stratum III, the moister, higher elevation middle part of the valley (Zeidler and Kennedy, 1994). In both Stratum I and Stratum III there

Figure 5.3 Traditional cultivation of maize and yuca: a. field interplanted with maize (sides) and yuca (center); b. ridgetop maize planting

are extensive deposits of river alluvium favored for CCP. Such lands are lacking in Stratum II. We knew that Stratum I and Stratum III were extensively settled in prehistory, with large settlements with ceremonial mounds occurring in each. The agricultural data, then, were collected to represent each stratum of the study area, and in the analysis, I consider a third hypothesis: that differences in agricultural productivity among the strata might explain the early intensive settlement of the San Isidro area in Stratum III.

We collected three types of maize data for the yield study: ear length, potential yield in ears (count) per hectare (ha), and recall yield data, also in ears/ha (see explanation below). For yuca, data on root weight per *mata* (plant), and measured yield in *matas* per hectare, were collected. For each field visit, we described the location (including slope and aspect) and located the field on an aerial photograph or topographic map. We asked the field or crop owners about cultivation practices (crop spacing and mix; timing of planting and harvest; use of fertilizer, herbicide, or insecticide), their assessment of the growing season and harvest, and past uses of the land. A total of 183 field visits were made.

How well a maize crop did over the course of a growing season can be evaluated in a number of ways. One way is to look at how large the ears are, that is, the length of the entire developed ear. To evaluate ear size, we measured a random selection of harvested ears, 20 if possible, from each field studied (Figure 5.4). From these measurements, I calculated average length of maize from the field. In the Jama Valley, maize for domestic use is stored "on the cob" in well-ventilated, elevated corncribs. In many cases, I measured stored ears, but if the maize was still in the field, the farmer picked the sample for me. In addition to length, I counted the number of rows, and noted the variety of maize (as identified by the farmer). Since yuca is only harvested as needed, it was difficult to obtain yield data (root weight per plant) for more than three or four individual plants. We dug out plants selected by the owner, and weighed the roots (Figure 5.5).

We kept back a few ears from most maize fields studied early in the project in order to establish the relationship, noted by Kirkby (1973) for traditional maize agriculture in Oaxaca, Mexico, between ear length and dry kernel weight for local varieties. By measuring ear length for 20 cobs in the field, mean ear length can be used to predict mean kernel weight for the crop of each field without having to shell and weigh the maize.

For fields with standing maize plants, I decided to measure the *potential* yield of the field by determining the density of the maize planting (number of ears per square meter), and converting this into ears/hectare. Planting density was quite variable, depending on whether yuca or other crops were interplanted with the maize, what the preferred planting distance was, and the number of kernels planted in each *mata,* or plant group (Figure 5.6). We measured distances between plants and rows, and counted numbers of plants in matas and numbers of ears on plants for these calculations. I used this approach during the first years of the project. Potential yield of yuca was measured the same way.

The last field seasons of the project, I gathered maize yield data after the harvest was completed by asking farmers for the total yield (*recall* yield data). I made this change because it had proven difficult to time fieldwork to correspond to the harvest. In 1988, for example, we were too early for many fields; in 1989, the fieldwork could not be done until summer, and most maize was harvested. Farmers readily recalled the most recent harvests, including the details of planting. The harvest data was usually given in *fanegas* (1 fanega is 1,800 ears), field size in *cuadras* (1 cuadra is 7,056 m^2, or 84 × 84 m), and planting distances in *varas* (1 vara is 84 cm). I converted these data to ears/ha.

As I mentioned earlier, cattle raising and plantation crops such as coffee, cacao, and banana (CCP) dominate the Jama Valley today. Lands that are likely the most productive—old river alluvium, groundwater-fed forests, and humid uplands—are

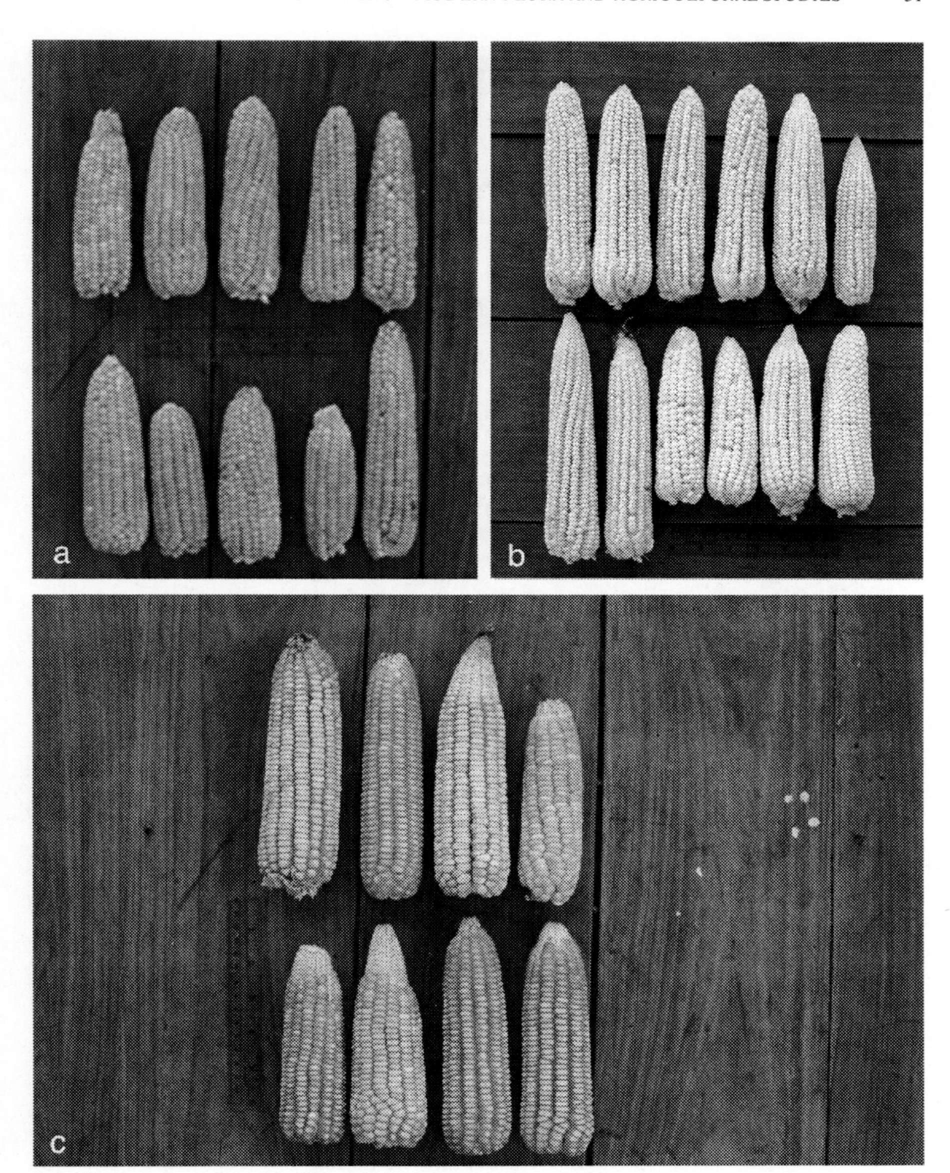

Figure 5.4 Some typical examples of the three varieties of maize commonly grown in the valley and studied during the project (the same 6" scale is in each photograph): a. amarillo, *an orange-yellow maize with soft, floury kernels; b.* cubano blanco*; a white to cream-colored maize with kernels of variable hardness; c.* hibrido *(hybrid 515 and 526), a yellow to orange hybrid maize with hard kernels*

planted in CCP, not maize and yuca. Maize plantings are tucked in on the edges of cash-crop plantings in these areas, but relatively few were available for study in comparison to less well watered fields (Figure 5.7). I reasoned that mapping the occurrence of CCP—crops very demanding of water and fertile soils—would serve as a

Figure 5.5 Harvesting yuca: a. plant ready for harvest; b. yuca roots attached to the base of the stem

proxy indicator of the location of lands that were potentially among the highest yielding for maize, but that could not be studied for maize and yuca productivity today.

Working on horseback and on foot with local guides, I sketched modern land-use patterns on aerial photograph overlays, as illustrated in Figure 5.7. This served to update and "ground truth" the patterns as indicated on topographic maps of the

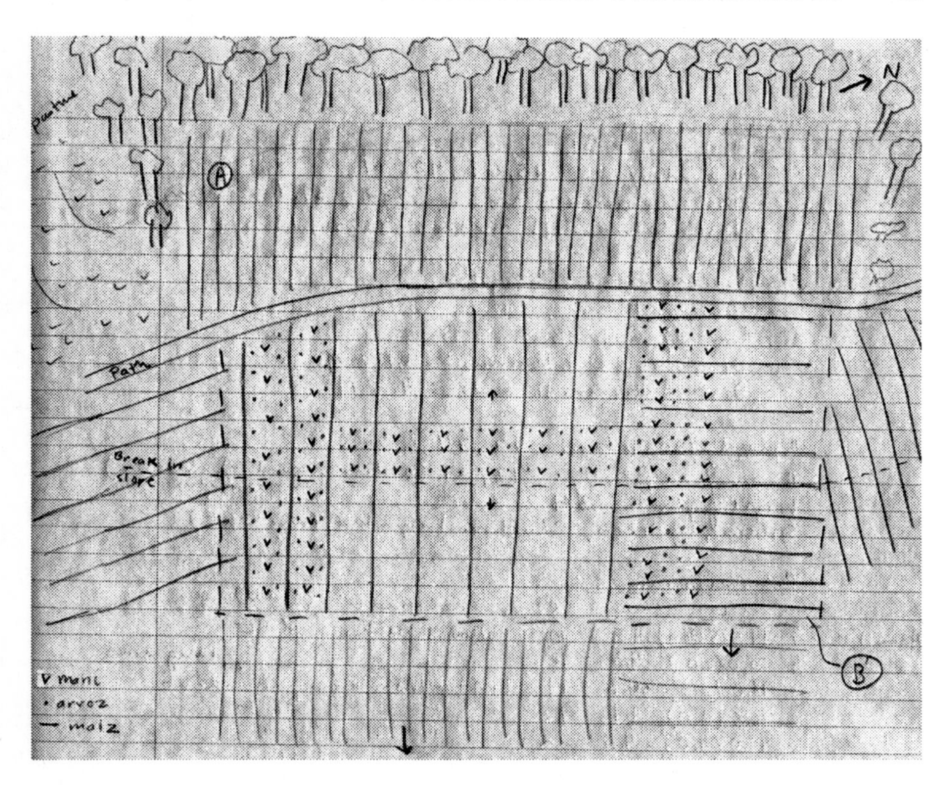

Figure 5.6 Sketch map of a field planted in maní (peanut), arroz (rice), and maíz (maize). Notice the difference in spacing between rows in sector A of the field (planted only in maize) and those in sector B (rows of maize alternating with plantings of rice and peanut). v = maní; • = arroz; — = maíz

region. Many areas marked as forest, for example, were found actually to be coffee or cacao plantations, while others had become cleared pastures.

• • •

PART II: SUMMARY

In Part II, Fieldwork, I have described the Jama River Valley and what it was like to live and work there, and reviewed the methods we employed to recover the archaeobotanical data and to understand how agricultural productivity varied around the valley.

There was, of course, much more to the fieldwork of the Jama Archaeological-Paleoethnobotanical Project than what I have reviewed here. The interested reader should consult *Regional Archaeology in Northern Manabí, Ecuador, Volume 1. Environment, Cultural Chronology, and Prehistoric Subsistence in the Jama River Valley* (Zeidler and Pearsall, 1994) for more details. In addition to the activities and results described there, an extensive pedestrian survey, covering 100% of the alluvial lands of the study region and randomly selected quadrats of the nonalluvial

Figure 5.7 Land-use map of the San Isidro area, showing the relationship of CCP (black lines), maize (white lines), and pasture (gray) plantings. Forest remnants are outlined in white. This map was made by sketching in fields on an acetate overlay on an aerial photograph. The best agricultural lands are in CCP.

uplands, was designed and carried out by Jim Zeidler over several long field seasons. The analysis of these data, still ongoing, in conjunction with the results from excavations at San Isidro and other selected sites, will allow us to test our models for the emergence and evolution of complex societies in northern Manabí province.

As a step to this larger goal, I will now move on to my analysis of plant-people interrelationships in the Jama River Valley. In the process, we will take a close look at the Jama-Coaque tradition.

6/Plants Used by the Jama-Coaque People

HOW PLANT REMAINS WERE IDENTIFIED

As I discussed in Chapter 4, charred plant remains and phytoliths, plant opal silica bodies, form the basis of our understanding of plants used by the prehistoric inhabitants of the Jama Valley. Before I discuss plants identified during the project, a few words about how I identified these remains are in order.

Identifying fragmentary remains of seeds, fruits, wood, tubers, and other macroremains requires direct comparisons of unknown material to known comparative specimens (Pearsall, 2000). This is a requirement for all paleoethnobotanical research, but is especially important in the tropics, where species diversity is high. In other words, there are more taxa to compare to unknown specimens in the tropics than in the temperate zones. In the case of wood, for example, the University of Missouri (MU) comparative collection for eastern North America contains some 65 specimens; the collection for coastal Ecuador, a much smaller area geographically, contains more than 150, and this is not a complete record of the woody flora. High arboreal diversity has led to difficulties in determining the identity of wood remains recovered in test excavations, a point to which I will return later.

A good laboratory comparative collection makes it much easier to identify fragmentary remains, since one can char the material, break it up, and study the appearance of the fragments. It is important to realize, however, that fragmentary archaeobotanical remains may not be identifiable beyond a very general level—the botanical family, for instance—because diagnostic characteristics are missing from the pieces, or because many of the plants in the family produce similar seeds or wood. One may also simply lack the "right" comparative specimen. The high species diversity of the lowland tropics necessitates a conservative approach to identification.

How then, exactly, are identifications made? Sources such as Pearsall (2000) and Hastorf and Popper (1988) provide more detail, but in brief the process is as follows. First, charred plant remains are removed from the modern debris (rootlets and the like) in the light fraction of flotation samples. At the MU lab, this step is carried out by student research assistants. Samples are first passed through a 2.0-mm geological sieve. Using a forceps, and working under low magnification (7–10 ×), assistants remove all charred materials from the > 2-mm fraction and sort these into broad categories: wood, porous material, dense material, rind/nutshell, maize kernels and cob fragments (cupules), large seeds, and other. Working under higher magnification (10–20 ×), smaller charred seeds and maize fragments are then removed from the <2-mm fraction, using either a forceps or a small paint brush (static electricity holds the seeds on the brush, as shown in Figure 1.3), and sorted into like categories. Finally, any nonbuoyant charred remains removed from the > 1-mm heavy fraction in the field lab are added to the light fraction materials.

Preliminary identifications and tallies of remains are now made. Identification actually starts during sorting, since the student assistants who carried out this first step readily learned the more easily recognizable remains, such as *Trianthema portulacastrum* and Solanaceae seeds, and maize cupules and kernels (Figure 6.1). As I reviewed the sorted samples (by year, initially, as the project progressed) and recognized other seed types, I taught these to the sorters. "Good types" (i.e., entire seeds with some distinguishing surface features) that I did not know were assigned an unknown seed number and sketched on an index card. Identified and unknown seeds from the < 2-mm fraction were counted and the counts entered on the form for the sample. As is common, identifications were often made at the level of the genus (e.g., *Passiflora*) or family (e.g., Solanaceae), and more rarely at the species level (e.g., *Trianthema portulacastrum*), since macroremains often lose the features, as a result of charring and abrasion by soil, needed for species-level identifications.

The same identification procedure was carried out on the > 2-mm material. I reviewed the separation of wood from fruit and tuber materials, and identified large seeds, such as cotton, recovered in this fraction. One summer following the final field season, I tackled the problem of separating the porous, or bubbly-looking, tissues of fragmented cotton seeds, maize kernels, and tubers from each other. I cut up charred comparative examples of all these, as well as a number of wild tree fruits collected during the project. I then mounted the various parts of each comparative specimen (epidermis, inner walls of the seed or fruit, endosperm or cotyledon tissue, tuber tissue with and without epidermis, and so on) on glass microscope slides with nail polish, and studied the morphology of the tissues. Finally, I summarized my findings on how the tissues differed, and, with the help of undergraduate research assistant Midori Lee, separated "porous" archaeological maize kernel fragments from broken cotton seeds, and root/tuber tissue from seed and fruit tissue. Also as part of preliminary identifications, I separated the variety of fragments of larger fruits encountered in samples into morpho-types, that is, "kinds" of things defined by shape and size features.

The final steps in identifying the macroremains, which I carried out just before this writing, involved going through each flotation sample one more time, to check identifications and the consistency of morpho-types, and to try to identify unknowns (unks). The first step is largely self-explanatory. I looked at all the materials in each sample again, and made sure all seeds labeled as *Passiflora* were *Passiflora,* that all the material tallied as "thick rind with internal divisions" looked the same, that beans (*Phaseolus vulgaris*) and Sapotaceae cotyledons were correctly distinguished, and so on for all the types.

The second step, identifying the unks, falls somewhere between searching for a needle in a haystack and solving a "whodunit." For some families or orders of plants, seeds and fruits share common characteristics. If an unknown looks a bit like one

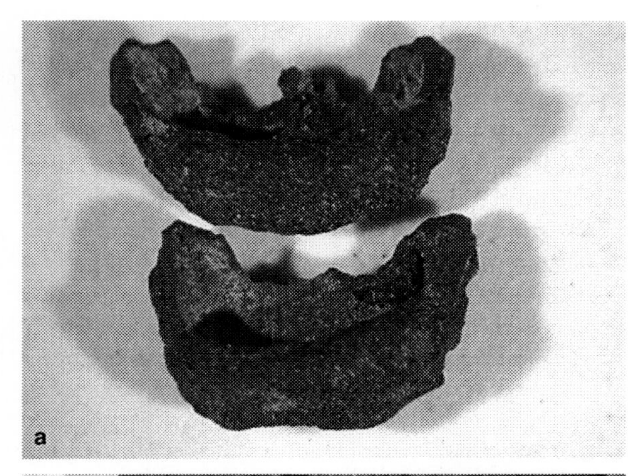

Figure 6.1 Charred maize from the Pechichal site in the Jama River Valley: a. cupules (average size 4.7 mm [width], 2.0 mm [length]); b. kernels (average size 6.1 mm [length], 7.6 mm [width], 4.2 mm [thickness])

genus, but the match is not exact, a paleoethnobotanist can sometimes identify the unknown by checking the other genera in the family, and then other families in the order. In the Jama material, for example, I had several kinds of small seeds that were flattish, circular in outline, with curved embryos, and had a fairly featureless seed coat. One type fit *Amaranthus* (Amaranthaceae) quite well, but another type was too small, and a third had a distinctive seed attachment scar that did not fit this family. I checked seven other genera in the Amaranthaceae (pigweed family) documented on the coast of Ecuador, and found that the smaller unknown was a fairly good match to the genus *Iresine*. I tallied these as cf. ("looks like") *Iresine*. I then checked two related families, Chenopodiaceae (lambsquarter) and Phytolaccaceae (pokeweed), for the larger unknown seed. Only one genus in the former occurs on the Ecuadorian coast, and it was not a good match. I checked six genera in the Phytolaccaceae and found my unknown: it was identical to seeds produced by the genera *Rivina, Hilleria,* and *Phytolacca*. I therefore made the identification at the family level: Phytolaccaceae. My searching for unknown seeds was carried out by using the MU paleoethnobotany lab collections, and by examining specimens curated at the MU Herbarium and the Missouri Botanical Garden Herbarium.

In the case of some unks, however, I had no idea to what family or order the unknown might belong. This was a problem, especially, with the highly fragmented remains of larger fruits. To work on these, I reviewed ethnobotanies of lowland South American indigenous peoples such as that by Balée (1994), and compiled a table of documented edible, wild tree fruits for which related species grew on the coast of Ecuador. With the permission of MU Herbarium director Robin Kennedy, I obtained specimens from the herbarium of many genera and families on this list that were not in the paleoethnobotany lab comparative collection, and charred and studied these, along with tree fruits collected during the project. I also studied the cultivated fruits I had purchased in markets in Latin America over the years, and reviewed family-level characteristics of fruits from published sources. These efforts allowed me to eliminate some ideas I'd had for unknowns, to separate palm nutshell more accurately from the thick rinds of other (unknown) fruits, and to make a few identifications; for example, I identified a fragment of a capsule (a type of fruit) with closely spaced spines, and an unknown seed type, as the dye/spice plant *achiote, Bixa orellana.* I matched disarticulated seed attachments that looked like small caps to seeds of *Sideroxylon,* a genus related to cultivated *sapote*. For the most part, however, the identities of the highly fragmented tree fruits in the Jama samples remain a mystery.

Identified macroremains and described morpho-types were tallied by sample, by site, using the computer spreadsheet Excel. An example of a complete data set, that of the Muchique 2 phase Pechichal site, is given in Appendix A. Table 6.1 summarizes the occurrence of macroremains at this site, our best data set for early Jama-Coaque II.

Phytoliths provide the other source of information on plants used during prehistory in the Jama River Valley. Phytoliths also provide insight into the vegetation of the valley in the past, and how human activities affected the environment.

Phytoliths are extracted from archaeological sediments and comparative soil samples through chemical processing that breaks the bonds between the microscopic silica bodies and soil or sediment constituents, then floats phytoliths out of the matrix (Pearsall, 2000). Specially trained student research assistants carried out this work. Extracted phytoliths are studied by mounting a small amount of extract (0.001 g) in Canada balsam on a microscope slide. After the slide "sets up" for a few days, the individual silica bodies can be rotated and examined from all sides under high magnification (250–400 ×) using a transmitted-light research microscope.

Phytoliths from archaeological samples are identified in the same way as macroremains, by direct comparison to modern plant specimens. Early in the Jama project, we scanned a number of slides of phytoliths from archaeological contexts using a process called *quick-scanning*. In this method, a slide is scanned row-by-row, and a list made of all the kinds of phytoliths encountered. Phytoliths are not counted; the slide is examined until redundancy is achieved (i.e., until nothing new is encountered on the slide). After quick-scanning a few slides, one has a pretty good idea of what will be encountered in samples, and can begin working on identifying which plants produced the forms.

Many of the identifiable, or diagnostic, phytoliths encountered in the Jama samples were known prior to our work, identified either during earlier research at MU or through the research of Dolores Piperno, a phytolith analyst based at the Smithsonian Tropical Research Institute in Panama (e.g., Piperno, 1988, 1989; Piperno et al., 2000). (The dry Pacific coastal forest of Panama has many similarities to the coastal Ecuadorian forest, making Piperno's research of great value to our

TABLE 6.1 MACROREMAINS FROM PECHICHAL

Site number	M3B4-011
Liters floated	834
Wood Ct.	2903
Cultivated	
Maize kernel fragment, Ct.	1762
Maize cupule fragment, Ct.	2340
cf. Manihot esculenta, Ct.	818
Monocot rhizome, Ct.	2
Root/tuber fragment, Ct.	590
Gossypium, Ct.	363
Phaseolus fragments, Ct.	651
Lagenaria rind, Ct.	8
Arboreal	
Aiphanes, Ct.	1
Phytelephas, Ct.	38
Arecaceae fragment, Ct.	228
Bixa orellana, Ct.	112
Mimosaceae (tree legume), Ct.	3
Psidium, Ct.	5
Sapotaceae seed fragment, Ct.	53
cf. Sideroxylon, Ct.	353
Thick rind, Ct.	197
Thick flat rind, Ct.	87
Thick rind, internal div, Ct.	44
Thin rind, Ct.	24
Thin rind w/storied str., Ct.	24
Curved rind, Ct.	3
Dense cotyledon fragment, Ct.	481
Small dense spherical fruit, Ct.	19
Dimpled fruit peel, Ct.	1
Unk 82	1
Unk 83	2
Unk 87, Ct.	2
Unk 104, Ct.	41
Porous endosperm fragments, Ct.	181

Small seeds	
Amaranthus (Amaranthaceae)	43
Phytolaccaceae	175
cf. Iresine	22
Asteraceae	1
Fabaceae (744)	1
Fabaceae, beaked	2
Fabaceae, elongated	1
Fabaceae, small	9
Fabaceae, medium	4
Fabaceae, large	2
cf. Linum (Linaceae)	59
cf. Abutilon (Malvaceae)	4
cf. Herissantia (Malvaceae)	94
Sida (Malvaceae)	12
Passiflora (Passifloraceae)	19
Poaceae, small-medium elongate	55
Portulaca (Portulacaceae)	6
cf. Rubus (Rosaceae)	3
Solanaceae	109
Teucrium	4
Trianthema portulacastrum	188
Unk 6	1
Unk 13	1
Unk 18	1
Unk 27	248
Unk 29	1
Unk 30	2
Unk 32	3
Unk 33	21
Unk 34	2
Unk 35	1
Unk 51	2
Unk 55	1
Unk 59a	2
Unk 85	6
Unk 86	5
Unk 91	1
Unk 93	1
Unk 97	1

project.) We also processed leaf and fruit specimens collected during the Jama project to extract phytoliths, and this yielded new types that were recognized in the archaeological samples (Figure 6.2). Cesar Veintimilla (1998, 2000) conducted much of the study of comparative specimens while at MU working on a master's degree, and he also scanned most of the archaeological samples.

A counting form, listing the plant taxa whose diagnostic phytoliths were encountered in the quick-scannned samples, was then designed for the project. Diagnostics included a variety of open-area indicators, arboreal forms, and economic plants. Figure 6.3, the phytolith diagram for a large bell-shaped pit excavated at the Pechichal site, illustrates what plants were identified by the presence of diagnostic phytoliths in samples. Each slide was examined until 500 diagnostic phytoliths were counted, or until the entire slide was examined. This method—counting to a fixed sum—is called *relative counting*. Occurrence of individual plants is thus expressed as a percentage of the sum. In Figure 6.3, percentages are read at the bottom of the diagram for each plant type listed across the top of the diagram. The plants identified in each sample—in this case, six trash deposits, or elements, in the Pechichal "Big Pit" (Feature 5)—are read horizontally, across the diagram from left to right. Samples are ordered stratigraphically, top (most recent) to bottom (oldest).

Most of the phytolith samples for the Jama project were scanned and counted in the early 1990s. Since that time, we have studied many more comparative specimens, and are able to identify a number of important plants more precisely than was possible before. (For more information about comparative phytolith studies at MU, see our Web site: http://www.missouri.edu/~phyto). For example, early in the project, we identified the family Marantaceae by the presence of nodular spheres. Since that time, research at MU, as well as research by Piperno (1989), has resulted in genus- and species-level identifications within this family of tropical root crops. Phytolith analysts can now identify cultivated arrowroot (*Maranta arundinacea*) and *llerén* (*Calathea allouia*), as well as wild species in these genera. It is now also possible to identify maize not only by cross-shaped leaf bodies, the method developed by myself and Piperno in the 1970s and 1980s (see Pearsall, 2000, for a history of the development of the maize identification method), but also by cob bodies (Pearsall et al., 2003). To "update" the Jama phytolith study, I asked research assistant Karol Chandler-Ezell to quick-scan a selected number of samples, usually one per cultural phase per site, for the presence of important new diagnostics. Presence of these is indicated by + on Figure 6.3.

PLANTS USED BY THE JAMA-COAQUE PEOPLE

As a result of the identification of macroremains and phytoliths in samples from 14 sites in the Jama River Valley, some 67 plant taxa (exclusive of wood) have been identified, and an additional 14 morpho-types and 39 unknowns described. Many of these taxa were encountered in flotation and phytolith samples taken from the Pechichal site (see Table 6.1, Figure 6.3, and Appendix A), and most occur in early Jama-Coaque II contexts. I will now describe the identified plants and discuss how they may have been used, and what they tell us about past land-use practices and environments in the Jama River Valley. Identified plants are presented by botanical family in three groups: economic plants, arboreal plants, and open-area/weedy plants. This structure reflects the interpretive framework I will use in Chapter 7 to

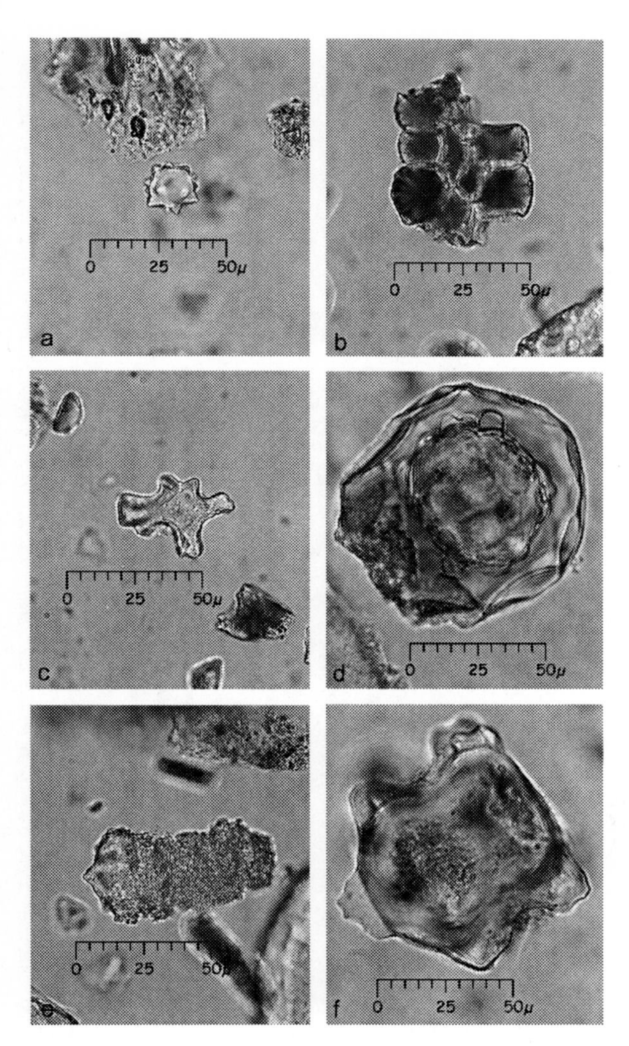

Figure 6.2 Diagnostic phytoliths observed in Jama samples: a. Arecaceae (palm) sphere; b. group of Cannaceae (achira) spheres; c. Dichapetalaceae epidermal body; d. Lagenaria siceraria (gourd) sphere; e. Marantaceae seed phytolith; f. Maranta (arrowroot) seed phytolith. See also Figure 1.4

discuss Jama-Coaque foodways. Detailed descriptions of all the archaeological types may be obtained by contacting me.

A note on how the following discussions are structured. For each entry, the first line lists the kind of material recovered (i.e., phytoliths or type of macroremain). Then I discuss the identification: at what taxonomic level it is placed, whether the taxon is domesticated or wild, what species occur in coastal Ecuador, and so on. Finally, I provide information on possible uses of the plant, and for domesticated taxa, summarize what we know about the history of the crop, and how it is grown.

I consulted a variety of sources to compile the information presented below. Gentry's (1993) field guide to woody plants, volumes in the *Flora of Ecuador* series (Harling and Sparre, 1973–1998), Jørgensen and León-Yánez's (1999) catalog of vascular plants, Little and Dixon's (1969) description of common trees of Esmeraldas Province, Acosta-Solis's (1961) description of the forests of Ecuador, Valverde's (1974) guide to orders, families, and genera of dicotyledons, and the

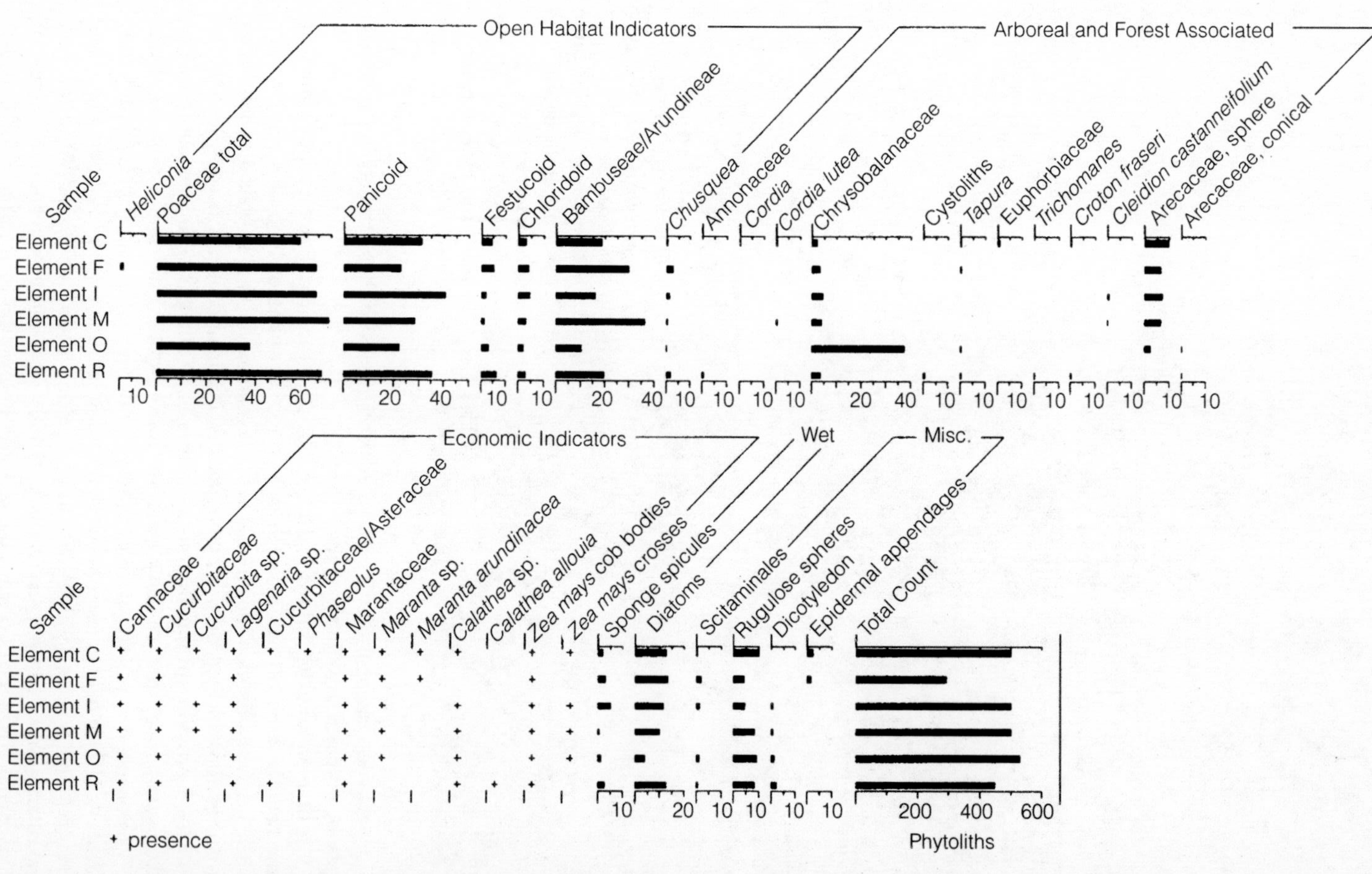

Figure 6.3 Pechichal resolved phytolith diagram. Elements (deposits) are in descending stratigraphic order. Phytoliths are groupd by ecological zone or type (open habitat, arboreal, economic, wet habitat, and miscellaneous types). The percentage of each phytolith in an element is read using the scale at the bottom of the graph. Presence of types observed outside the count is indicated by +.

report by Neill and colleagues (Neill et al., 1999) on botanical exploration in the Mache-Chindul mountains all contributed to my understanding of the geographic distribution and habitat preferences of species discussed here, as did fieldwork conducted by myself and Robin Kennedy (Zeidler and Kennedy, 1994) in the Jama Valley. For additional understanding of which species might be expected in open or disturbed habitats (i.e., weeds), I consulted Ferreyra (1970) and all my plant-collecting notes for such habitats for coastal Ecuador. I found a number of ethnobotanies, field guides, and handbooks useful for determining which wild plants were edible, had medicinal properties, or were useful in other ways. Principal among these was Balée's (1994) in-depth study of plant utilization by the Ka'apor; I also consulted Anderson and Posey (1989), Boom (1989), Castner et al. (1998), Duke (1992), and Vickers and Plowman (1984). Finally, for information on economic plants, including likely areas of origin and cultivation requirements, I relied on Piperno and Pearsall (1998), Purseglove (1968, 1972), Simmonds (1976), and Smith et al. (1992), all works focused on tropical crops.

Economic Plants

Into this category fall plants that were probably cultivated—planted, tended, or otherwise cared for—by the prehistoric inhabitants of the Jama River Valley. Not all species of economic value in the lowland tropics are cultivated or have undergone the process of genetic change that marks domesticated plants, of course. For the purposes of discussing plant-people interrelationships in the Jama Valley, however, it seems useful to consider as a group those plants that are typically planted and cared for in prepared fields or house gardens, and that have documented economic value.

CANNACEAE (*CANNA* FAMILY)—*Canna* (*achira*) phytoliths (Figure 6.2b) *Canna* is a genus of large herbs that grow in moist, open settings. A number of species have edible rhizomes (underground stems). One of these, cultivated *achira*, *Canna tuerckheimii* (= *C. edulis*) is a possible source of *Canna* phytoliths in the Jama samples (Figure 6.4). A second native cultivated species with edible rhizomes, *C. indica*, also occurs in Manabí, however (escaped and established in secondary vegetation), and two native wild species (*C. glauca*, *C. jaegeriana*) are listed elsewhere on the Ecuadorian coast, although not in Manabí Province.

Edible, cultivated *achira* is no longer commonly grown outside the Apurimac region of Peru, but was once grown from the Antilles to Argentina and through the Amazon basin. It can be cultivated from sea level to 2000 m elevation. Although there is insufficient information to suggest a precise area of origin for *achira,* open habitats within seasonally dry forest at the fringes of the tropics are likely areas. *Achira* is grown by planting small, cormlike rhizome segments at the beginning of the rainy season. Approximately eight months is required for a crop to mature, but it can be left in the ground for up to two years. The sweet rhizomes can be baked, boiled, or eaten raw, or the starch can be extracted.

CUCURBITACEAE (SQUASH/GOURD FAMILY)—*Lagenaria siceraria* (bottle gourd) rind fragments, Cucurbitaceae fruit stalks, and Cucurbitaceae, *Cucurbita,* and *Lagenaria siceraria* phytoliths (Figure 6.2d) Truly wild populations of bottle gourd occur today only in South and East Africa, and it is considered native to the semidry

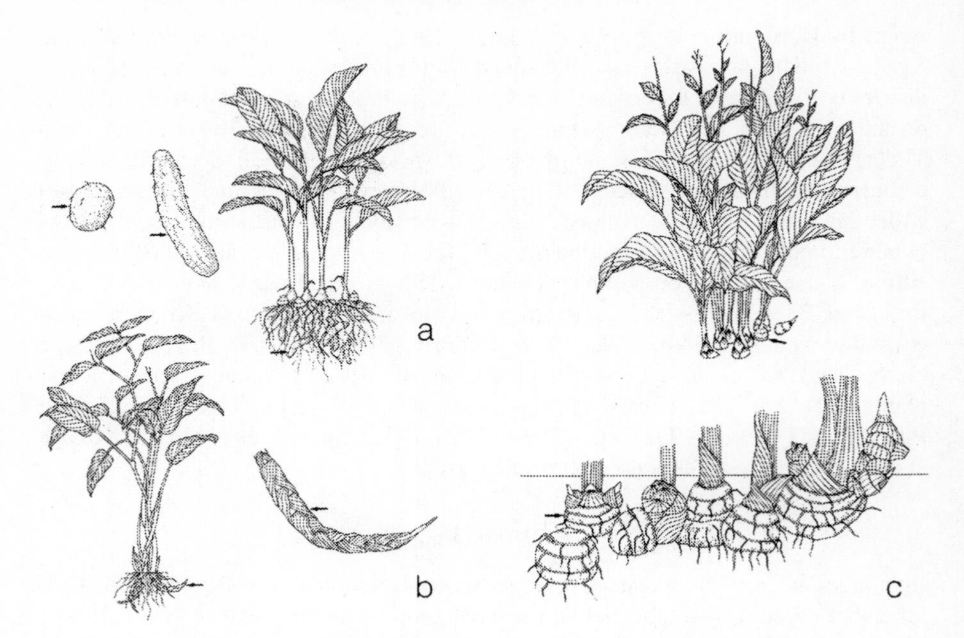

Figure 6.4 Root/tuber resources identified in Jama samples: a. Calathea allouia, llerén; b. Maranta arundinacea, arrowroot; c. Canna tuerckheimii, achira. From The Origins of Agriculture in the Lowland Neotropics, *Dolores R. Piperno and Deborah M. Pearsall, Fig. 3.3 and 3.5. Reproduced with permission from Academic Press.*

tropical lowlands of Africa south of the equator. It is likely that wild African gourds washed out to sea in the south Atlantic were carried west by the south equatorial current and ended up on the coast of Brazil or northern South America. Humans took over as dispersal agents and incorporated the bottle gourd into house gardens all around South America. The flesh of most gourd cultivars is too bitter to eat, but the oily seeds are edible. The primary use of the bottle gourd was probably as a container and net float, however. Charred rind fragments of bottle gourd, as well as robust spheroidal phytoliths produced in the rind, were recovered from samples.

Five native species of squash (*Cucurbita*) were brought under domestication in the New World (Figure 6.5). While squash of modern commerce are grown for their flesh, initial cultivation was likely for the edible seeds and usefulness of containers because the flesh of wild squash is stringy and bitter. Without going into detail on squash domestication, it is possible that wild squashes growing in southwest coastal Ecuador (to the south of the Jama study region) were locally domesticated and then "lost" when another lowland species was introduced. Robust spheroidal phytoliths of the type produced in rinds of domesticated squashes were recovered from samples, as were large silicified hairs common to both *Cucurbita* and *Lagenaria*. Squash and gourds are grown from seed, and are often interplanted with maize.

EUPHORBIACEAE (SPURGE FAMILY)—cf. *Manihot esculenta* (yuca) root fragments In the Big Pit at the Pechichal site are preserved several large pieces of charred root that resemble comparative specimens of yuca. The genus *Manihot* is represented by

K. Chandler-Ezell

Figure 6.5 Cucurbitaceae fruits: a. squashes, Cucurbita ssp.; b. gourd, Lagenaria siceraria

some 100 species, ranging from southern Arizona to Argentina, with most being native to arid or seasonally dry regions and open habitats. Many species produce enlarged roots with starchy reserves (Figure 5.5). An incredible amount of variation exists in cultivated yuca, some of which is likely the result of crossing among related species.

The domestication history of this important lowland crop is complex, but recent research makes a strong case for a South American origin, either in northern South America or central Brazil. Yuca roots are rich in starch and, pound per pound, provide a more efficient source of carbohydrates than any other New World root or tuber crop. Yuca is also high-yielding, even on poor soils, and very undemanding. (See Chapter 8 for a discussion of yuca cultivation.) Some strains are highly poisonous ("bitter"), and must be processed to remove glucosides before cooking. The yuca grown prehistorically in the Jama Valley was most likely a "sweet" variety: bitter manioc cultivation in South America is restricted today to the eastern lowlands.

FABACEAE (BEAN FAMILY)—*Phaseolus* (bean) seeds; *Phaseolus* hooked hair phytoliths Numerous entire halves (single cotyledons) of bean seeds and some whole beans were recovered from Jama samples (Figure 6.6). Entire half and whole beans were especially numerous at the Pechichal site. All specimens lack the seed coat. The size and shape of the cotyledons are consistent with a small variety of *Phaseolus vulgaris,* the common bean, rather than Lima bean (*P. lunatus*) or jack bean (*Canavalia* spp.). The whole beans from Pechichal (N = 16) range in size from 6.2–9.8 mm (long), 3.8–6.15 mm (wide), and 3.15–4.9 mm (thick). The average size is 8.6 mm × 5.2 mm × 4.0 mm. This population is quite similar in size to charred beans recovered from sites in Peru, and smaller than uncharred archaeological or modern beans. Shrinkage during carbonization occurs. Small hooked (recurved) silicified hairs produced in *Phaseolus* were observed in one deposit at the Pechichal site.

Wild common beans grow in relatively dry environments from Mexico to the southern Andes of Peru, Bolivia, and Argentina. In Ecuador, wild bean populations tend to be found on the western slopes of the Andes, above 1000 m elevation (i.e., to the east of the Jama Valley region). Multiple lines of evidence indicate that beans were domesticated two or three times from local populations of a widespread wild ancestor. Beans are an important source of protein and carbohydrates. First use may have been as a vegetable in the green stage (as we use green beans today), but after the development of ceramic containers, which permitted longer cooking times, the usefulness of mature, dried *Phaseolus* beans would have been realized. Beans are grown from seed. They can be grown on most soil types, from light sands to heavy clays, and today are often interplanted with maize on good, alluvial soils.

MALVACEAE (COTTON FAMILY)—*Gossypium* (cotton) seeds Four species of cotton (Figure 6.7) are known from Ecuador. Two are endemic to the Galápagos Islands; one is native to and cultivated in Guayas Province (*Gossypium hirsutum*), and one is native to and cultivated widely on the coast (including Manabí), Andes, and Amazon (*G. barbadense*). Based on the present-day distribution, the Jama seeds are likely *Gossypium barbadense.*

The two domesticated cotton species have a complex history, both before and after domestication. Briefly, *Gossypium barbadense* likely emerged as a species, and was brought under domestication, in the coastal plain of northern Peru/southern Ecuador. *G. hirsutum* emerged as a species either in the west coast of southern Mexico or coastal northern Peru/southern Ecuador. The domestication of the latter species probably occurred after it spread as a wild plant into coastal regions of northern South America and the Caribbean, however. Cotton is a sun-loving plant that

Figure 6.6 Charred beans, Phaseolus vulgaris, from the Pechichal site

does not thrive under shade, heavy rainfall, or cool conditions. Grown as an annual crop today, traditional cotton varieties are perennial shrubs. The cotton fiber (actually seed hairs or lint) is contained in a fruit that splits open when ripe. Natural color ranges from white to cream to chocolate brown. Cotton can be twined to create nets, carrying bags, and hammocks, and can be woven into cloth.

MARANTACEAE (ARROWROOT FAMILY)—Marantaceae phytoliths, *Calathea* and *Calathea allouia (llerén)* phytoliths, *Maranta* and *Maranta arundinacea* (arrowroot) phytoliths (Figure 6.2e, f) The Marantaceae are a family of robust herbs that often produce tubers. Four genera are known from the coast of Ecuador, including Manabí: *Calathea, Ischnosiphon, Maranta,* and *Thalia.* In addition to cultivated *llerén, Calathea allouia,* six wild *Calathea* species occur on the coast; in addition to cultivated arrowroot, *Maranta arundinacea, M. gibba* also occurs. *Calathea* and *Maranta* species favor lowland forest understory habitats, and may commonly be found in seasonally moist sites, such as along streams or ponds. In secondary vegetation, naturalized *M. arundinacea* is common along roadside ditches.

Calathea allouia, llerén, is cultivated in the Caribbean region and in northern South America for its edible tubers (Figure 6.4). *Maranta arundinacea,* is also considered indigenous to northern South America and the Caribbean. It produces starchy rhizomes that are very tough, requiring thorough grinding or maceration to release the starch. Insufficient information is known about wild related species of either crop to designate an area of origin. It is also quite possible that other tuber-producing species of *Calathea* and *Maranta* were grown in the past. Arrowroot grows best on rich, sandy loam soil. It cannot tolerate waterlogging. Propagation is by rhizome tips; because these are commonly broken off during harvest of the rhizomes, the crop often self-propagates. Planting occurs at the beginning of the rainy season, and mature rhizomes are produced in 10–12 months. *Llerén* is reproduced from the rootstock, since the tubers lack eyes. *Llerén* tuber production requires about 12 months. The starch of both arrowroot and *llerén* is of high nutrient value and easily digestible.

Figure 6.7 Modern cotton seed, Gossypium barbadense (right), compared to a charred example from Pechichal (left)

MONOCOTYLEDON GROUP—Monocot rhizome A few pieces of charred rhizome exhibiting a pattern of nodes similar to a small *Canna* or sedge rhizome were recovered from Jama samples. Another specimen, lacking a complete diameter, has a folded epidermis with several nodes. These finds are thus rhizomes (origin in stem tissue), rather than roots. Phytoliths from three rhizome-producing monocot crops occur in Jama assemblages: *Canna, Maranta,* and *Calathea.* Sedge (Cyperaceae) is also documented in some phytolith samples.

POACEAE (FORMERLY GRAMINEAE; GRASS FAMILY)—*Zea mays* (maize) cupule and kernel fragments and *Zea mays* leaf and cob phytoliths (Figures 6.1, 6.8) The story of the origin and evolution of maize is a complex one that has been the topic of many books and articles. Briefly, the most widely held position is that maize originated from teosinte—wild *Zea* native to southern Mexico and Guatemala—and was introduced by humans into Central and South America, on the one hand, and into America north of Mexico on the other. Both these dispersals happened during prehistory, but the spread southward is the more ancient. The early maize spread by tropical forest agriculturalists south out of Mesoamerica was characterized by small, hard kernels that would not have been a highly desirable starch source. Eventually selection favored increased kernel size, however, and the thickness of the kernel epidermis thinned, and changes in endosperm hardness occurred, leading to increases in productivity and ease of use of maize.

Today there are many varieties of maize with flint, flour, dent, and sweet endosperm, as well as various kinds of popcorn, the hardest type of endosperm. In addition to providing carbohydrates, maize is a source of protein, a better source than many root crops. After kernels are dry, maize can be stored without any processing, although it is subject to insect damage. I describe maize cultivation practices in Chapter 8.

What kind of maize was grown in the Jama River Valley in prehistory? There are several difficulties inherent in answering this question. Maize is a very diverse crop, with hundreds of varieties existing today. It is likely that even more kinds of maize

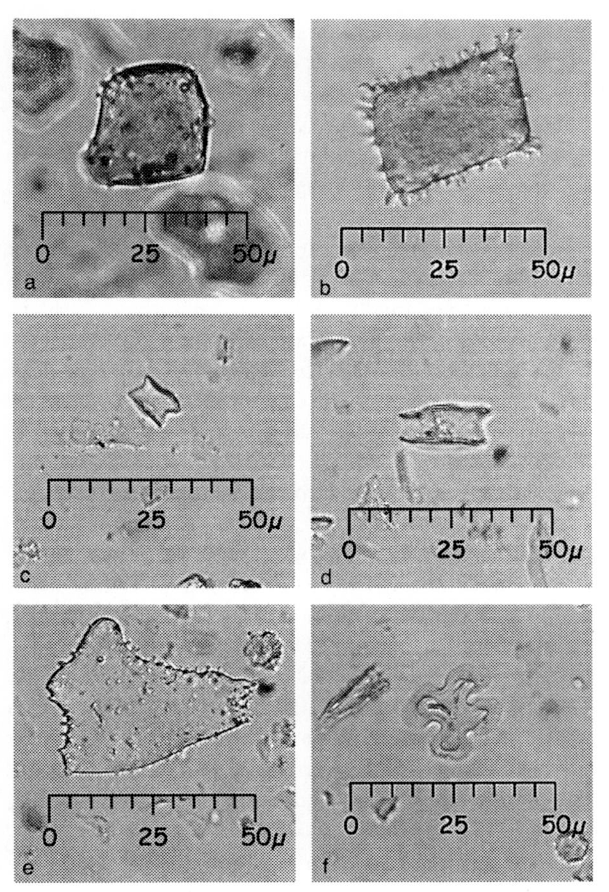

Figure 6.8 Maize phytoliths from Jama samples:
a. globular cob body;
b. wide IRP (irregular body with projections) cob body;
c. wavy-top rondel cob body;
d. ruffle-top rondel cob body;
e. irregular IRP cob body;
f. Variant 1 maize leaf body

were grown in the recent past, before the spread of hybrid, improved varieties, and the loss of traditional ones. This great diversity makes it challenging to compare archaeological maize to all similar varieties. Furthermore, there is no reason to assume that one will find a match—the kind of maize grown in the Jama Valley 1,500 or 3,000 years ago may not exist today. As I discussed in Chapter 3, there is no cultural continuity between prehistoric and present-day inhabitants of the valley; traditional Jama-Coaque maize varieties may have been lost when human populations perished. Finally, even though charred maize remains are relatively abundant in Muchique 2 sites such as Pechichal and El Tape (see Chapter 7), they are highly fragmented. No entire maize cobs were recovered. This increases the difficulty of characterizing the kind of maize that was used.

What I can say about maize used in Muchique 2 times is based on study of individual cupules—structures of the cob that hold the kernels in place—and kernels (Figure 6.1). I measured width, length, and edge angle of 40 cupules from the Pechichal site in the middle valley (randomly selected from 4 flotation samples). Width averages 4.7 mm, length 2.0 mm, and edge angle 44 degrees. Size ranges are 3.1–6.65 mm (L), 1.4–2.35 mm (W), and 30–70 degrees (angle). While the cupules are variable in size, they are quite uniform in shape (e.g., in proportion of length to width, depth, shape of the edges or "wings"). These findings suggest to me that one type of maize was deposited at Pechichal. The average cob had 16 rows of kernels (a 44-degree angle indicates 8 cupules would fit around the cob, with 2 kernels seated in each), and was slender and probably not too long. There are only 2 whole cupules from the El Tape site in the lower valley; their measurements fall close to the average for Pechichal (4.4 mm wide, 2.0 mm long [both cupules], 50-degree angle [average]).

There were many fewer whole kernels preserved at Pechichal, 13 in all. These average 6.1 mm in length and 7.6 mm in width, are 4.2 mm thick, and have an average edge angle of 28 degrees. Angles are much harder to measure on kernels, because the tissue "puffs" during charring, but this average is a pretty good fit to the cupule data (kernel angles of 22.5 degrees would be expected for 16-row maize). Most of the charred maize from El Tape is kernels, but few are complete. Many are just "caps," the top of the kernel, so it is impossible to measure length or angle. Average width is 6.6 mm; average thickness 4.3 mm. I compared the Pechichal and El Tape kernel measurements by Student's *t*-test, and found no significant size difference between them.

It appears likely, then, that a small-kerneled, somewhat variable 16-row maize was grown throughout the valley in Muchique 2 times. If I correct the measurements of the archaeological kernels to account for swelling during charring (Pearsall, 1980), average sizes would be 5.5 mm (L), 7.2 mm (W), and 2.8 mm (T). This is smaller than kernel sizes of the two existing traditional coastal varieties *(Candela [Amarillo]* and *Chococeño),* and differently proportioned than existing maize from eastern lowland Ecuador (Timothy et al., 1963). This Jama-Coaque maize may have died out, or evolved into one of the existing lowland varieties. Its kernels were not as large as those of maize grown in the valley today, and its ears were likely quite a bit smaller in diameter and length.

We can also learn something about the kind of maize used in the valley by looking at maize cob phytolith assemblages in samples from the Pechichal Big Pit (Figure 6.8). Table 6.2 illustrates the frequency of various cob bodies recovered from two sediment samples, as well as from a sample of charred maize cupules from the pit. Without going into detail on these results, the Pechichal maize produced an assemblage of inflorescence phytoliths observed in modern maize samples, except for the occurrence of two forms—half-decorated oblong and robust globular—that are more often observed in teosinte, the ancestor of maize (Pearsall et al., 2003). The Pechichal maize is also characterized by more abundant short-cell phytoliths (rondels) than long-cells (bodies with speculate projections and other forms) in the cobs, a pattern that characterizes many modern maize varieties. This archaeological variety thus shows an interesting blend of primitive (teosinte-like) and derived phytolith traits, but due to the large numbers of rondels has an overall "advanced" look.

The observant reader has probably already noticed that the assemblage of cob bodies from the sediment samples of the Big Pit does not exactly match the phytoliths extracted from charred cupules from the same deposits. There are many fewer ruffle-top rondels in the sediment samples than in the cupules, for example. While these differences could be an outcome of charring, or differences in procedures for extracting phytoliths from sediments and charred tissues, I suspect that the more

TABLE 6.2 MAIZE COB PHYTOLITHS AND RELATED FORMS FROM PECHICHAL*

	Rondels			Bodies with Speculate Projections							Other		
	wavy-top	ruffle-top	½-decorated	rectangular IRP, wide	rectangular IRP, narrow	irregular IRP	robust globular	gracile globular sphere	tubular	½ dec. oblong	semicut	burrlike	Total
Charred Cupule Extract													
#	8	29	0	9	1	4	11	2	0	4	0	0	68
%	11.8	42.6	0	13.2	1.5	5.9	16.2	2.9	0	5.9	0	0	
Big Pit #749													
#	6	6	0	17	1	15	35	0	0	0	0	0	80
%	7.5	7.5	0.0	21.3	1.3	1.9	43.8	0	0	0	0	0	
Big Pit #751													
#	9	13	0	22	0	9	21	0	1	5	0	0	80
%	11.3	16.3	0	27.5	0	11.3	26.3	0	1.3	6.3	0	0	

*Recovered from charred cupules and two sediment samples from Feature 5 (Big Pit).

fragile forms like ruffle-tops were broken in the rough and tumble of the burial environment of the Big Pit.

MORPHO-TYPE—Root/tuber fragments This morpho-type is comprised of fragments of underground storage organs that lack features allowing further identification. It is likely that small pieces of yuca root are included here—I suspect much of the root/tuber material from Pechichal is yuca, since material closely resembling that root is preserved in the Big Pit—as well as fragments of monocot rhizomes such as *Canna, Maranta,* and *Calathea.* However, all fragments lack an epidermis, and thus important features such as presence or absence of leaf nodes and root scars, or the arrangement of nodes, are not available to permit more precise identification. Most pieces are also too small and distorted by charring to discern patterning in the vascular elements. I have determined that these are fragments of roots or tubers by comparing the material to roots and tubers in the lab collection.

Arboreal Plants

Into this category fall the many and diverse kinds of tree fruits encountered in the Jama samples. It is impossible to determine from the archaeobotanical remains of these plants which taxa were planted in house gardens, which were left standing when agricultural fields were cleared from forest, and which were gathered from wild stands of trees. By Jama-Coaque II times, there was little land in the Jama

Valley that was not affected by human activities, making it likely that useful trees were maintained by people one way or another.

ANNONACEAE (SOURSOP FAMILY)—*Annona* (soursop) seed fragments and Annonaceae phytoliths There are 25 genera and 106 species of Annonaceae in Ecuador, including some 13 genera native to the lowland coastal forests (Figure 6.9). In broad terms, the occurrence of Annonaceae phytoliths can be considered an indicator of moist lowland forest. Some species also grow in secondary forest. Various species of *Annona* are cultivated for their juicy or meaty edible fruits. Fruits of *A. muricata, guanábana,* for example, can be eaten raw and used to make fruit drinks. Medicinal uses are also reported. *Guanábana* is easily and rapidly grown from seed. *Rollinia mucosa* and *Guatteria microcarpa* also have edible fruits. A few members of the family produce wood useful for construction or furniture making.

ARECACEAE (FORMERLY PALMAE; PALM FAMILY)—*Aiphanes* (*coroso* palm) fragments, *Phytelephas* (*cadi* palm) fragments, Arecaceae seed fragments, and Arecaceae phytoliths (Figure 6.2a) A few palm fruit fragments were complete enough to be recognizable as *coroso* palm. *Aiphanes eggersii* is the most common species of *coroso* in Manabí, but *A. tricuspidata* also occurs. Fragments of *cadi* palm—producer of *tagua* nuts, or vegetable ivory—were readily identifiable (Figure 6.10). *Phytelephas aequatorialis* is the only species listed for Manabí province. While *tagua* nuts are used today for producing carvings, beads, and other items for sale, immature seeds are edible. Charred palm remains identifiable only at the family level include robust seed coat fragments and endosperm fragments. *Bactris* seeds resemble the archaeological fragments in thickness and curvature; *Bactris* and *Astrocaryum* are among the genera that produce endosperm of the type found in the samples. Palm phytoliths were common in all samples.

Palms are among the most useful "trees" of the tropical forest, and are discussed in most ethnographic accounts of plant use in lowland South America. The large fronds are valued as roofing material and for making mats, for example (Figure 6.11). Leaf fibers can be used for bowstrings and for weaving hammocks, clothing, carrying straps, and ropes. Fleshy seed coats of some species are edible, while for others the endosperm is eaten or processed to extract an edible oil. Arrow shafts may be made from the leaf petiole. Fruit-eating larvae can be harvested from some species. Roots and "woody" leaf petioles can be used to produce household items such as yuca graters or loom parts. Palm wood is used in construction and as bow wood.

Palms are a conspicuous part today of the arboreal flora of the Jama River Valley. They are often left at the edges of cleared fields and in house gardens (i.e., Figure 3.4a). *Cadi* palm was planted widely in the mid-20th century for the vegetable ivory it produces. Introduced *palma real,* the African oil palm (*Roystorea*), is common throughout the valley, as are coconut plantings along the sea shore. *Coroso, cadi,* and *chonta* palm (*Bactris* spp.) are among the most common native species in the valley. For Manabí as a whole, some 19 species of palm have been documented in 13 genera.

Figure 6.9 Wild Annona tree with a single fruit

Figure 6.10 Carved and entire tagua nuts, vegetable ivory from the cadi palm (Phytelephas), with a piece of the spiny fruit

Figure 6.11 Palm frond roof on a small house. The walls are made of split and flattened bamboo canes.

BIGNONIACEAE (TREE GOURD FAMILY)—cf. *Crescentia cujete* (tree gourd) rind fragments Tree gourds—not related botanically to squash or bottle gourd—are a fairly common component of the drier coastal forest today, and also occur in the Andes and Amazon of Ecuador. This species of small- to medium-sized trees is described as introduced and cultivated, and is naturalized on the coast. Extensively cultivated through most of tropical America, the native range is unknown, but tree gourd probably originated in northern Central America. Tree gourd fruits are more or less spherical with a hard, woody shell. They can be used as containers or decoratively carved for sale.

BIXACEAE (*ACHIOTE* FAMILY)—*Bixa orellana (achiote)* fragments Remains of *achiote* preserved in Jama samples include a single fragment of the fruit—a spiny capsule—and seeds. *Bixa orellana* is native and cultivated on the coast of Ecuador; it is the only species of the genus on the coast, and is often naturalized in secondary forest. *Achiote* is widely cultivated in the tropics; its range today extends from Mexico to Bolivia. It occurs as a shrub or small tree, growing 2–10 m tall. Annato dye, used as a food colorant, dye, and as a body paint, is obtained from the pulp around the seeds by macerating them in water.

BORAGINACEAE (*BORAGE* FAMILY)—*Cordia, Cordia lutea (mullullu)*, and *Cordia alliodora (laurel)* phytoliths All species tested in the genus *Cordia* (Figure 1.4b) deposit abundant silica in their epidermal tissues, producing distinctively shaped trichomes (prickles and hairs) and trichome bases. *Cordia* is a genus of trees, shrubs, and vines with 15 representatives on the Ecuadorian coast. *C. lutea* (*mullullu*) and *C. macrocephala* are listed for Manabí province, *C. alliodora* (*laurel*) has been collected in the Jama Valley, and other species might be expected. *Mullullu,* a small tree that attains 7 m in height, is a fairly common component of secondary growth and open habitats in the drier coastal formation (Figure 6.13d). *Mullullu* is cultivated as an ornamental tree (it produces an abundance of clear, yellow flowers), its fruits produce a gum that can be used as a glue, and the wood is gathered for fuel. *Laurel,* a medium to large tree that grows to 45 m, has been documented by Kennedy as a major invader species in Strata II and III (Zeidler and Kennedy, 1994). *Laurel* occurs in the greatest number of random survey quadrats in both strata, followed by *guasmo (guácimo), Guazuma ulmifolia* (Sterculiaceae), discussed below. *Laurel* produces a high-quality wood used in construction and furniture making, and is planted for its wood and as a shade tree in coffee plantations. Seeds and leaves are used medicinally. Occurrence of various *Cordia* species phytoliths thus provides evidence for the presence of open, secondary forests in the valley.

BURSERACEAE (*PALO SANTO* FAMILY)—Burseraceae phytoliths The Burseraceae is a family of arboreal, and some shrubby, species. The genera *Bursera, Dacryodes, Protium, Tetragastis,* and *Trattinnickia* occur on the Ecuadorian coast. Only *Bursera graveolens* (*palo santo,* occurring as a shrub or tree) is listed for Manabí province, but a number of species in other genera occur to the north, in Esmeraldas, and may well extend into Manabí. Many members of the Burseraceae have aromatic resin or gum in their tissues, including wood. *Palo santo* wood, for example, is burned as an incense or as a smudge fire to drive away insects. Medicinal and ritual uses are also noted. Some species have durable wood. In broad terms, the occurrence of Burseraceae phytoliths can be considered a forest indicator.

CHRYSOBALANACEAE FAMILY—Chrysobalanaceae phytoliths The Chrysobalanaceae is predominantly a family of arboreal species. The genera *Hirtella, Licania,* and *Parinari* occur on the Ecuadorian coast; several species of *Hirtella* and *Licania* are listed for Esmeraldas, as is *Parinari romeroi.* Unpublished results of recent collecting trips by Neill and colleagues (1999) place *Hirtella mutisii, Licania celiae,* and *L. licaniiflora* in northern Manabí. *Licania* species are usually trees of the upper forest canopy. Occurrence of Chrysobalanaceae phytoliths thus indicates the presence of moist, mature primary forest or forest remnants in the Jama Valley (Figure 6.12).

DICHAPETALACEAE FAMILY—cf. *Tapura* phytoliths (Figure 6.2c) This family of trees, shrubs, and vines is represented in Ecuador by three genera, *Dichapetalum, Stephanopodium,* and *Tapura. Tapura* is only listed for the Ecuadorian Amazon, however (an Esmeraldas collection of *Tapura angulata* has been reclassified as *Stephanopodium*). *Dichapetalum* and *Stephanopodium,* both with coastal species, are thus more likely sources for the Jama phytoliths, but have not yet been tested. In broad terms, cf. *Tapura* phytoliths can be considered a forest indicator.

EUPHORBIACEAE (SPURGE FAMILY)—*Croton, Cleidion,* and Euphorbiaceae phytoliths The Euphorbiaceae are a large family of herbs, shrubs, and trees, represented in Ecuador by 50 genera and 243 species. The genus *Croton* contains 39 species, 10 of which occur on the coast. *Croton* species are typically shrubs or small trees that favor open habitats and forest clearings. *Cleidion castanneifolium* is the only species of its genus in Ecuador. It is a small tree or shrub that occurs commonly in the understory of moist forests.

Figure 6.12 Characteristic vegetation of the moister forest

HYMENOPHYLLACEAE (FILMY FERN FAMILY)—*Trichomanes* phytoliths Trichomanes is a genus of ferns that occurs in moist, forested habitats.

LAMIACEAE (FORMERLY LABIATAE, MINT FAMILY; *VITEX* TRADITIONALLY PLACED IN VERBENACEAE)—cf. *Vitex gigantea* (*pechiche*) fruit pit fragment *Pechiche* is a distinctive tree of the drier coastal forest formation, its presence marked by a rain of brillant blue flowers during the dry season. One fairly complete *pechiche* fruit pit was recovered, and it is likely some unidentified fruit fragments also come from this tree fruit. There are three native *Vitex* species on the Ecuadorian coast, *V. flavens, V. gigantea,* and *V. triflora*. The first two are large trees; the third is a shrub. Ripe *pechiche* fruits are eaten today on the coast of Ecuador, but I was unable to find other ethnographic accounts of uses of this genus.

MIMOSACEAE (TREE LEGUME FAMILY) Several seeds produced by a tree legume (Mimosaceae) were recovered in Jama samples. These seeds are similar to those produced by *Prosopis (algarrobo), Pithecellobium, Acacia, Mimosa,* and *Inga,* among other genera. Legume trees are a conspicuous component of the very dry tropical forest of Stratum I, and also occur in the more humid, inland areas of the valley (Figure 6.13). *Prosopis* and *Acacia* provide browse for animals and a hard, resinous wood that produces a hot-burning fuel. Fruit pods and seeds from several *Inga* species are documented as a usable, if not highly desirable, food source. One species, *Inga edulis,* produces a sweet pulp around the seeds.

MYRTACEAE (GUAVA FAMILY)—*Psidium* (guava, *guayaba*) seeds There is a good match between seeds from comparative specimens of *Psidium guajava,* cultivated *guayaba,* and the Jama archaeological specimens. However, one native *Psidium* also occurs in Manabí, *P. acutangulum. Psidium* is a large genus of small trees and shrubs, native to tropical America and the Caribbean, which has numerous species with edible fruits. Some species can be quite invasive. *P. guajava,* the cultivated *guayaba* or guava, is a shrub or small tree that bears yellow fruits, rich in vitamin C, with numerous seeds embedded in the sweet flesh. The flesh is mixed with water to create a fruit drink, or dried into a sweet paste. Guava is also used medicinally.

SAPOTACEAE (SAPOTE FAMILY)—cf. *Sideroxylon* (dilly) seed coat fragments Detached hilums (seed attachment features), shaped like small caps, are fairly abundant in Jama flotation samples. A few were recovered "in situ," that is, with fragments of robust seed coat surrounding the cap. These attachment features and seed coat fragments resemble specimens of *Sideroxylon* (= *Mastichodendron*) *foetidissimum* provided to the MU lab by L. Newsom. They do not resemble other comparative specimens in this family. It is likely, then, that the archaeological seed coat fragments are from the native Manabí tree *Sideroxylon obtusifolium,* but it is possible that an unexamined species in a related genus contributed these fragments to the assemblage. Wild species in the genera *Chrysophyllum, Microphilis, Pouteria,* and *Pradosia* occur in coastal Manabí and Esmeraldas provinces, and a number are cultivated for fruit, including *Pouteria caimito.*

Sapotaceae cotyledon fragments A few charred cotyledon halves too wide and too thick to be *Phaseolus* (bean) were recovered in Jama samples. These cotyledons were likely produced by one of the Sapotaceae tree taxa that occur on the coast. I was unable to determine whether these cotyledons match the seed coat fragments, however, since no cotyledon fragments were recovered with seed coats or had hilums attached to them.

Cultivated Sapotaceae tree fruits have sweet to mealy flesh embedded with one to several large seeds. The fruit pulp is eaten fresh or mixed with water to make a fruit drink. Fruit skins of a number of species contain an unpleasant-tasting latex. The trunk of *Pouteria sapota* (*sapodilla* or *chico zapote*) is tapped for latex that is made into a base for chewing gum (*chicle*). Use of wild fruits of *Chrysophyllum, Micropholis,* and *Pouteria* species as secondary food sources is documented for the Ka'apor of Brazil (Balée, 1994).

STERCULIACEAE (CHOCOLATE FAMILY)—*Guazuma (guázmo)* phytoliths These phytoliths are very similar to those produced by a comparative specimen of *Guazuma ulmifolia, guázmo. Guázmo* is a characteristic tree of dry areas, especially open areas and secondary growth. It is a morphologically variable small to medium tree that grows to 15 m. The wood is useful for household items, but is very susceptible to

Figure 6.13 Trees of the drier forest: a. Ceiba; b. Ochroma pyramidale (balsa); c. Prosopis (algarrobo)

Figure 6.13 (continued) d. Cordia. Balsa and some species of Cordia grow abundantly in disturbed dry habitats.

termite attack. The flowers are used medicinally. As mentioned earlier, *guázmo* is an invader of cleared areas in the Jama Valley. The Sterculiaceae is a family of small trees and shrubs with a number of coastal representatives. *G. ulmifolia* is the only species in its genus in Ecuador, and is widespread throughout the country.

ULMACEAE (ELM FAMILY)—*Celtis* (hackberry), *Trema*, and Ulmaceae phytoliths
The Ulmaceae is a small, arboreal family represented by three genera in coastal Ecuador, *Ampelocera, Celtis,* and *Trema*. *T. macrantha* is the only coastal species in that genus; this is a small to medium tree, 7–15 m, common in open sites and cleared fields, and typical of secondary forests. *Celtis* is represented by two coastal species, one of which, *C. shipii,* favors humid forest settings. I do not have ecological information on the three species of *Ampelocera* that occur on the coast.

MORPHO-TYPES The following arboreal morpho-types were identified: thick rind, thick flat rind, thick layered rind, thick rind with internal divisions, thin rind, thin rind with storied structure, curved rind, dense cotyledon fragments, small dense spherical seed, dimpled fruit peel (all charred macroremains), cystoliths, and sclerids (both phytoliths).

The 10 charred morpho-types are fragments of different kinds of large fruits or seeds, including single and multilayered seed coat fragments, cotyledon pieces, and fruit pericarp fragments. The most commonly occurring morpho-type is dense cotyledon. These are fragments of large cotyledons or hard endosperm from large seeds. While I cannot be certain that these 10 morpho-types pertain to arboreal species, it seems likely, given the size and robustness of the fragments.

Cystoliths are phytoliths that are hemispherical, spherical, or bulbous-elongated in shape, usually with verrucose to rough surfaces. A few families produce smooth cystoliths, however. Cystoliths are produced in a variety of families, including the Acanthaceae, Bombacaceae, Boraginaceae, Flacourtiaceae, Moraceae, and Urticaceae, all families with many shrub and tree species. They thus serve as generalized arboreal indicators. Silicified sclerenchyma cells, or sclerids, are elongated, irregularly angled phytoliths produced very widely in arboreal species. Like cystoliths, sclerids are a generalized arboreal indicator.

UNKNOWNS As described at the beginning of this chapter, distinctive-looking material that could not be readily identified was usually assigned a number, and tallied as an unknown. Some of these unks were eventually identified, others merged with the commonly occurring morpho-types just enumerated. Ten unknowns still remain that are probably tree or shrub fruit fragments (Unk 67, 69, 70, 73, 77, 82, 83, 87, 99, 104). I will not describe these, except to note that they are sufficiently different to remain as separate types. In other words, while there is no way to be sure that these are the remains of different tree fruits, it seems likely that they are. Most, if not all, of these are single occurrences. I believe these are fragments of tree fruits (rather than fruits or seeds of herbaceous plants) because of the size and robustness of the remains.

CHARRED WOOD As I mentioned earlier, it proved difficult to identify charred wood recovered during the project to species or genus, and I will not discuss wood use in any detail here. This difficulty was due to a combination of factors, including the great diversity of the woody flora, the highly fragmented nature of many specimens, and the difficulties Kennedy and I encountered in identifying comparative specimens collected without flowers or fruits.

In spite of these challenges, graduate assistant Peter Warnock and I characterized some 47 morpho-types of charred wood, with the assistance of an undergraduate student, Russell Gaines, who prepared comparative specimens and worked on initial identifications. Results to date are summarized in Table 6.3.

While identifications are far from precise for most morpho-types shown in Table 6.3, at least 20 kinds of wood/woody stems were recovered from Jama flotation samples that resemble arboreal species, large shrubs (Solanaceae), palms, or vines (Vitaceae) that grow in remnant forest stands in the valley today. The remaining morpho-types do not match specimens in the comparative collection. Tree legumes are well represented, with nine discrete types characterized. Many legume trees produce dense, resinous wood that preserves well and is distinctive in appearance. These factors may lead to overrepresentation of these wood types in samples. However, dense legume woods also produce hot-burning fires, and may also have been selected by the valley inhabitants for that reason, or because they were plentiful in the dry forest. *Guayacán,* cf. *Tabebuia,* is another dense wood that is favored today for charcoal production. Palm "wood," *cereza de monte, Psidium* (guava), and cf. *Sapotaceae* were perhaps incidental inclusions in cooking fires, since all produce more valuable products than fuel: edible fruits, and in the case of palm, thatching material.

Open-Area and Weedy Plants

This final category of plants includes wild species that thrive without a forest canopy, in naturally open areas, fallow agricultural fields, and around human habitation sites (Figure 6.14). Not all are weeds in the sense of being especially inva-

TABLE 6.3 DISCRETE WOOD MORPHO-TYPES REPRESENTED IN JAMA ARCHAEOLOGICAL SAMPLES[1]

Fabaceae (tree legumes, as broadly defined), 9 types

E1601, *seca* (Unk 3, Unk 45)
cf. *Platypodium* (Unk 6)
E1629, *huarango* (Unk 12, Unk 15)
E1642, *algarrobo* (Unk 19)
cf. *Mimosa* (Unk 20)
cf. *Inga* (Unk 38)
cf. Fabaceae (Unk 7)
cf. Fabaceae (Unk 21)
cf. Fabaceae (Unk 22)

Arecaceae (palm), 2 types

cf. *Aiphanes* (Unk 13)
Arecaceae

Other types

Bignoniaceae, cf. *Tabebuia* (Unk 1)
Malpighiaceae, E1621–23, 1692, *cereza de monte* (Unk 18)
Euphorbiaceae (Unk 47)
Myrtaceae, cf. *Psidium* (Unk 43)
Solanaceae (Unk 41)
Vitaceae (Unk 42)
Rhamnaceae, cf. *Piptadenia* (Unk 23)
cf. Sapotaceae (E744)
cf. E1663, lengua de vaca (Unk 10)

[1]Collection numbers and common names, when available (e.g., E1601, seca) are given for taxa matched to unidentified MU comparative specimens. Original morpho-type numbers are in parentheses (e.g., Unk 3)

sive or unwanted; on the contrary, some of these sun-loving plants were likely encouraged and used as greens, spices, or medicinal plants. This is one of the hardest categories of archaeobotanical remains to interpret, since it can be difficult to distinguish between seeds that were accidentally charred when blown into a cooking fire and seeds of a medicinal plant that were discarded in the fire after the greens were steeped for a medicinal tea. Similarly, phytoliths from open-area plants may represent either plants intentionally brought into the site to be used or the "background" assemblage of phytoliths in the soil of the site. To aid in creating this category and interpreting these data, I reviewed a number of sources on tropical weeds, including Ferreyra (1970), and tallied up all my personal observations of plants that favored open habitats. These observations were recorded in my plant pressing notebooks.

AIZOACEAE (CARPET WEED FAMILY)—*Trianthema portulacastrum* seeds This is one of the most common seed types recovered in Jama flotation samples. *Trianthema portulacastrum* is a native herb of the Ecuadorian coast, the only species in the genus. I have collected it in a variety of open habitats; it is especially common in the rainy season. The young greens are edible.

Figure 6.14 Open area and weedy plants: a. recent alluvium is commonly covered with plants favoring sunny, open habitats; b. a stand of sedge (Cyperus) along a dry stream bed

AMARANTHACEAE (PIGWEED FAMILY)—*Amaranthus* (amaranth, pigweed) seeds Six species of *Amaranthus* occur on the coast. All are annual or rarely perennial herbs that favor open habitats. I have collected *Amaranthus* specimens from roadsides, cleared fields, dry streambeds, and other open habitats, especially after the start of the rainy season. The young greens and seeds are edible.

Figure 6.14 (continued) c. Solanum (Solanaceae) and d. Hyptis (Lamiaceae) in a fallow field

cf. *Iresine* seeds Two species of *Iresine* occur widely on the Ecuadorian coast; both are herbs or shrubs. I have collected *Iresine* from a grazed, open forest setting.

Asteraceae (Formerly Compositae; Sunflower Family)—Asteraceae phytoliths and seeds This large family is well represented in the Ecuadorian flora, with some 217 genera and 918 species. I have collected Asteraceae in a variety of open settings on the coast, including cultivated fields, roadsides, cleared fields, seashore, dry streambeds, and along streams. Sunflower and daisy relatives are common components of the weedy, rainy season flora of the Ecuadorian coast. Seeds of some Asteraceae are eaten.

COMMELINACEAE (SPIDERWORT FAMILY)—*Commelina* seeds These very distinctive archaeological specimens are a good match to *Commelina;* no other genus in the family produces seeds of the same size and shape. Two species are documented on the Ecuadorian coast, *C. diffusa* and *C. erecta,* both herbs. I have collected *Commelina* from streamsides, cleared fields, and maize fields.

CYPERACEAE (SEDGE FAMILY)—Cyperaceae phytoliths The sedge family (Figure 1.4c) is an abundant phytolith producer. While it is possible to differentiate among the major genera of sedges, genus-level identifications were not made for the Jama samples. The Cyperaceae are represented by 28 genera and 217 species in Ecuador. Many species of *Cyperus* and *Eleocharis,* two of the larger genera, occur on the coast. Sedges favor moist, open habitats, and some species invade disturbed environments like agricultural fields (Figure 6.14b). I have collected sedge specimens on the banks of dry streams, in flooded areas, in harvested fields, on field walls, and along streams and canals. Some sedges produce edible rhizomes, and the stems and leaves can be used to weave household items such as mats, or watercraft like the *Scirpus* (*totora*) reed boats of Lake Titicaca in Peru.

FABACEAE (BEAN FAMILY)—Fabaceae seeds The bean, or legume, family is divided taxonomically into three subfamilies, which are considered separate families by some botanists. Many species in one subfamily/family, the Papilionoideae, produce beanlike seeds; for example, *Phaseolus* beans, described under economic plants, belong to this group. A number of other kinds of wild Fabaceae seeds were recovered in Jama flotation samples. These are: Type 744, beaked (one Ecuadorian genus that produces seeds of this shape is *Crotalaria*), elongated (*Sesbania* produces similar, but larger seeds), small (*Rhynchosia, Desmodium, Indigofera, Aeschynomene,* and *Dolichos* produce similar seeds), medium (*Vigna* seeds are similar), and large (these look like tiny, entire *Phaseolus* beans).

There are 77 genera of Fabaceae in Ecuador, making the process of genus-level identification very time-consuming. For that reason, and because little additional information would be gained, I did not attempt genus determinations. I have collected numerous wild Fabaceae on the coast from a variety of open habitats, including roadsides, edges of canals, field walls, maize fields, dry streambeds, streamsides, and cleared fields.

LAMIACEAE (FORMERLY LABIATAE; MINT FAMILY)—*Teucrium* seeds *Teucrium* is one of six genera of mints that occur in coastal Ecuador. The Jama archaeological type is probably *T. vesicarium,* the only species that occurs on the coast. Mints are common components of open vegetation, especially during the rainy season; I have collected various species from maize fields, roadsides, streamsides, and inundated areas. Mints are valued for the many uses of their aromatic foliage.

LINACEAE (FLAX FAMILY)—cf. *Linum* seeds This seed is one of the most common in the Jama macroremain assemblage. Referred to from the first year of lab work as Unk 11, it has proven difficult to identify. Illustrations of *Linum* (Linaceae) resemble the unknown, and a comparison to *Linum* species in the MU Herbarium resulted in a close match. This genus is only listed for the Andes and Galápagos, however, not the Ecuadorian coast. For now, I have decided to leave the identification as cf. *Linum,* and will continue to work on the identification. There is only one genus and five species in this small family, and with two Galápagos species documented, it is possible that *Linum* has simply been overlooked on the coast. The Galápagos species are all herbs or subshrubs.

MALVACEAE (COTTON FAMILY)—Malvaceae seeds Three types of Malvaceae seeds occur in Jama samples: cf. *Abutilon,* cf. *Herissantia,* and *Sida.* The first two are uncommon in the Jama samples, while *Sida* seeds are fairly common.

There are some 18 native Malvaceae genera on the coast of Ecuador, 8 species of which are in the lab comparative collection. I consulted these to arrive at the identifications of the three types in the Jama samples. While *Abutilon* and *Herissantia* remain provisional identifications at the genus level, there is no doubt in my mind that these are Malvaceae. I have collected numerous species in this family from open habitats, including open, grazed areas, streamsides, roadsides, cleared fields, and seashore localities.

MUSACEAE (BANANA FAMILY)—*Heliconia* (bird-of-paradise) phytoliths The Musaceae is a family of rhizotamous plants, often very tall and robust, and mostly herbaceous. *Heliconia* is represented by five species in Manabí, with others listed for other coastal provinces. *Heliconia* species share a preference for moist, open habitats, and are often associated with openings in mature forests (Figure 6.15). *Heliconia* moves into light gaps and recently disturbed forest vegetation.

PASSIFLORACEAE (PASSION FRUIT FAMILY)—*Passiflora* (passion fruit) seeds Several Jama flotation samples contained uncharred (i.e., modern) passion fruit seeds, and a few charred seeds were recovered as well. The charred seeds match wild *Passiflora* species in the comparative collection, and are like miniature cultivated passion fruit seeds. The 91 species of *Passiflora* in Ecuador include many coastal representatives. I have collected wild *Passiflora* vines from maize fields and cleared fields. The two major cultivated species, *P. edulis* (passion fruit) and *P. quadrangularis* (giant *granadilla*), are grown throughout the tropics for their edible fruits. The aromatic pulp is mostly used for making juice, but can also be eaten raw or used to flavor jellies, jams, ice cream, and sherbet. Use of fruits from a number of wild species—including *P. foetida,* a common coastal weed—is also documented.

PHYTOLACCACEAE (POKEWEED FAMILY)—Phytolaccaceae seeds I described earlier in this chapter how I identified these large seeds as Phytolaccaceae, probably from the genera *Rivinia, Hilleria,* or *Phytolacca.* The Phytolaccaceae is a family of erect or clambering herbs or shrubs. *Rivinia humilis* occurs in Manabí; this perennial herb is widespread in the tropics. *Hilleria secunda,* an erect herb, also occurs in the province, as do three species of *Phytolacca.* The young shoots and leaves of *P. americana,* the North American pokeweed, are edible. I have collected *Phytolacca* along roadsides.

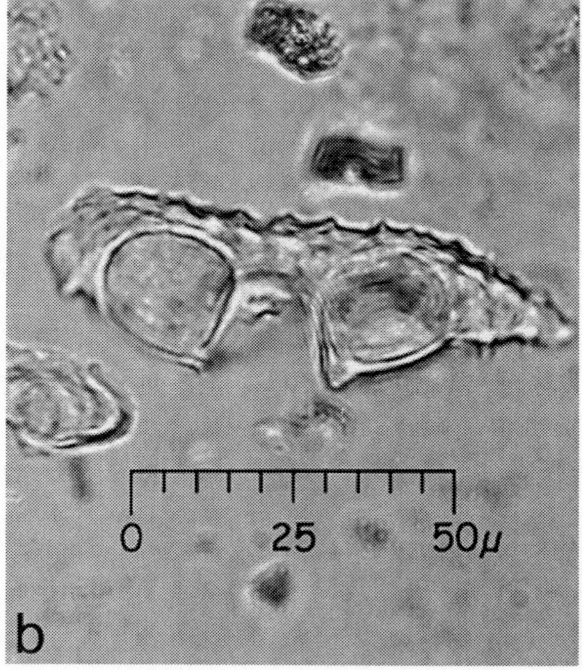

Figure 6.15 Heliconia: a. plant; b. diagnostic phytoliths

POACEAE (FORMERLY GRAMINEAE; GRASS FAMILY)—Seeds and phytoliths Evidence for grasses is common in archaeological contexts sampled during the Jama project, especially those sampled for phytolith analysis. Grasses are prodigious silica accumulators; all Jama phytolith samples contain abundant grass short-cells, a type of phytolith produced in the area of the leaf vein and in leaf-derived tissues. Charred grass seeds are less common, but nonetheless present in some flotation samples. Grasses dominate open habitats on the coast, especially during the rainy season, and are tenacious invaders of agricultural fields. A few species, such as bamboos, occur in forested habitats (Figure 6.16).

I identified one distinctive charred grass seed, cf. *Setaria,* and defined two other macroremain morpho-types, small-medium elongate grass and robust grass.

Panicoid, festucoid, chloridoid, Bambuseae/Arundinae (bamboo/cane), and *Chusquea* (bamboo) phytoliths were identified in samples from Jama sites. Many grasses that produce panicoid phytoliths favor moist, warm environments, and panicoid short-cells were usually abundant in Jama samples. In addition, maize is a panicoid phytolith producer. Many grasses that produce festucoid phytoliths favor moist, cool environments, while grasses that produce chloridoid phytoliths typically favor dry, warm environments. These two classes of grass phytoliths were present, but uncommon, in Jama samples.

Bambuseae/Arundinae (bamboo/cane) phytoliths occurred frequently in samples. Bamboos and canes grow today along watercourses in the Jama Valley (see Figure 6.16) and would be readily incorporated into phytolith samples on alluvial sites. In addition, bamboo is very useful as a building material and for making household objects (see Figure 6.11). *Chusquea* (bamboo) was also identified. This genus of bamboos favors forested environments. Most species are Andean, but *Chusquea simpliciflora* occurs in Manabí.

POLYGONACEAE (KNOTWEED FAMILY)—*Polygonum* (knotweed) seeds Five native *Polygonum* species occur on the Ecuadorian coast. I have collected *Polygonum* specimens along the edges of canals and in agricultural fields. Greens and seeds of some species are eaten.

PORTULACACEAE (PURSLANE FAMILY)—*Portulaca* (purslane) seeds There are five native *Portulaca* species on the coast. I have collected specimens in a variety of open habitats, including edges of canals, agricultural fields, field walls, roadsides, and banks of dry streams. Young leaves and stems of purslane are edible.

ROSACEAE (ROSE FAMILY)—cf. *Rubus* (raspberry) seeds Several examples of seeds resembling *Rubus* were encountered in Jama flotation samples. All are fragmentary, however, and so the identification is provisional. There are two native *Rubus* species on the Ecuadorian coast, *R. boliviensis* and *R. urticifolius,* both shrubs. I do not know if these produce edible fruits, but a number of members in the genus do.

SOLANACEAE (NIGHTSHADE FAMILY)—Solanaceae seeds The Solanaceae are the source of many culturally useful plants, including the edible potato, tomato, and eggplant; the spice chile pepper; the stimulant tobacco; as well as several poisons, including the deadly nightshade and jimson weed. The family is also abundantly represented in the weedy flora of the tropics by numerous species of *Solanum* and

Figure 6.16 Bamboo—plants with narrow stalks and fluffy leaves—growing along the edge of a small stream in the middle valley

Physalis (Figure 6.14c). Seeds of these genera (nightshade and ground-cherry) are very difficult to tell apart, and there are many species of each in the Ecuadorian coastal flora. The Solanaceae seeds in the Jama flotation samples are probably *Solanum* and/or *Physalis*. However, since I have not checked all 35 genera that occur in Ecuador, I will leave the identification at the family level. (The seeds are definitely not chile pepper, tobacco, or jimson weed.) I have collected *Physalis, Solanum,* and other members of this family in open habitats such as roadsides, stream edges, seashore, agricultural fields, dry streambeds, and field walls. Fruits of ground-cherries and *some* nightshades are edible.

MORPHO-TYPE—Small diameter stems As the name implies, this morpho-type consists of robust, charred stems that are not woody.

UNKNOWN TYPES Seeds that could not be readily identified during sorting or the preliminary identification stage were assigned numbers, as described earlier for arboreal unknowns. While I was eventually able to identify some of these, 29 seed types remain unknown (Unk 2, 6, 7, 13, 18, 24, 27, 29, 30, 32, 33, 34, 35, 39, 40, 49, 50, 51, 55, 58, 59, 59a, 80, 81, 85, 86, 91, 97, 100). Most of these are single seed occurrences; many seeds are broken and probably unidentifiable. While I cannot be certain that each type pertains to a different species of plant, it is likely that many do. I have classified these unknowns as open-area plants because the seeds are small and relatively fragile, like the seeds of many herbaceous plants.

Other Macroremains and Phytoliths

BROMELIACEAE (BROMELIAD FAMILY)—Bromeliaceae phytoliths Bromeliads produce very small, spinulose spheres that are also produced by palms. Since this type is likely a mix of the two families, it is not easily interpreted.

SCITAMINALES (ZINGIBERALES ORDER)—Scitaminales phytoliths This order of robust herbs, incorporating the families Zingiberaceae, Cannaceae, and Marantaceae, produces (among other forms) spherical phytoliths with folded/angled surfaces. Although the Scitaminales is mostly a group that favors moist habitats, both forest-dwelling and more open-area taxa are included in it. This order-level phytolith type is thus not very useful for interpreting past environmental conditions in Manabí. (As an aside, occurrence of folded/angled spheres in dry, coastal settings [not the normal habitat for this group] is associated with cultivation of *achira, Canna tuerckheimii.*)

RUGULOSE SPHERICAL PHYTOLITHS A difficulty similar to that just described occurs with interpreting this morpho-type. Rugulose spheres occur in the Cannaceae, Marantaceae, and Musaceae, among other families. It is difficult to use this type for interpreting past environmental conditions, since both forest-dwelling and open-area plants are included in these families.

DICOTYLEDON PHYTOLITHS We no longer count these phytoliths as diagnostics, but did so during the Jama project. These are the polyhedral and anticlinal (nonquadrilateral) epidermal cells produced by a variety of dicotyledon taxa, both woody and herbaceous. This type is of little diagnostic value.

SILICIFIED EPIDERMAL APPENDAGES Similarly, this category of silicified hairs and prickles is of limited diagnostic value, since it contains both grasses and dicotyledon epidermal appendages.

POROUS ENDOSPERM FRAGMENTS This is a fairly common macroremain morphotype, comprised of small bits of tissue with large, thin-walled cells (i.e., porous in appearance) and no other distinctive features. I place the morpho-type here, under "other," because it is difficult to determine whether these fragments were produced by arboreal or herbaceous taxa. I suspect that much of this material is small bits of maize and cotton endosperm tissue, but any number of other seeds with abundant starchy or fleshy endosperm could contribute to this type.

FUNGAL TISSUE Fungal spores as well as tissue from fungus fruiting structures were present in many Jama samples. It is likely that this material was introduced on firewood, and thus preserved by charring.

SPONGE SPICULES AND DIATOMS Diatoms and some types of sponges have silica skeletons that preserve in sediments and soils, like phytoliths produced by higher plants. In general, both classes favor moist habitats.

7/Plant-People Interrelationships in Early Jama-Coaque II

What do the plants recovered from sites in the Jama River Valley tell us about plant-people interrelationships during the first phase of Jama-Coaque II, the Muchique 2 phase (A.D. 400–750)? As I discussed in Chapter 2, this is the phase for which we have the most well-preserved sample of charred macroremains, as well as abundant phytolith data. Further, prehistoric populations in the valley, including those at the regional center, San Isidro, adapted successfully to the environmental impacts of the Tephra II ash fall event. What was the nature of the subsistence system that permitted this society to survive and prosper in the face of a disaster that had led to valley depopulation in the past?

At the time of this writing, five sites tested during the Jama project have deposits affiliated to Muchique 2 (Table 4.2). These are M3B3-002, El Tape; M3B4-011, Pechichal; M3D2-001, San Isidro; M3D2-009, Finca Cueva; and M3D2-056, Dislabón. Refer to Figure 4.1 for site locations. All the sites are located on broad alluvial terraces of the Jama River or one of its major tributaries.

I will not repeat the site settings described in Chapter 4; however, it is important to review the contexts of the samples, since this has some bearing on the results of the study. At El Tape in Stratum I, four test units were opened and sampled for macroremains and phytoliths. Feature 4, an early Muchique 2 floor, was extensively sampled. Deposits affiliated with the Formative period and the Muchique 1 phase were also encountered in the test excavations.

Excavations at Pechichal, in the western part of Stratum III, were limited to salvaging a bell-shaped pit—Feature 5, the "Big Pit"—which was eroding from a river cutbank (Figure 7.1). The natural stratigraphy of the pit consisted of a series of fill episodes referred to as elements and numbered A–T. Charred plant remains were visibly abundant in many elements. Radiocarbon determinations and analysis of ceramics place the fill of the Big Pit and the overlying deposits in the Muchique 2 phase, and suggest that the pit was filled with debris quite rapidly.

The paleoethnobotanical sample from the 40-ha, multicomponent San Isidro site, located on a major left-hand tributary of the Jama River, comes from six excavation areas: Sector V, Sector XII, Sector XVIII, Sector XX, Sector XXI, and Sector XXXI.

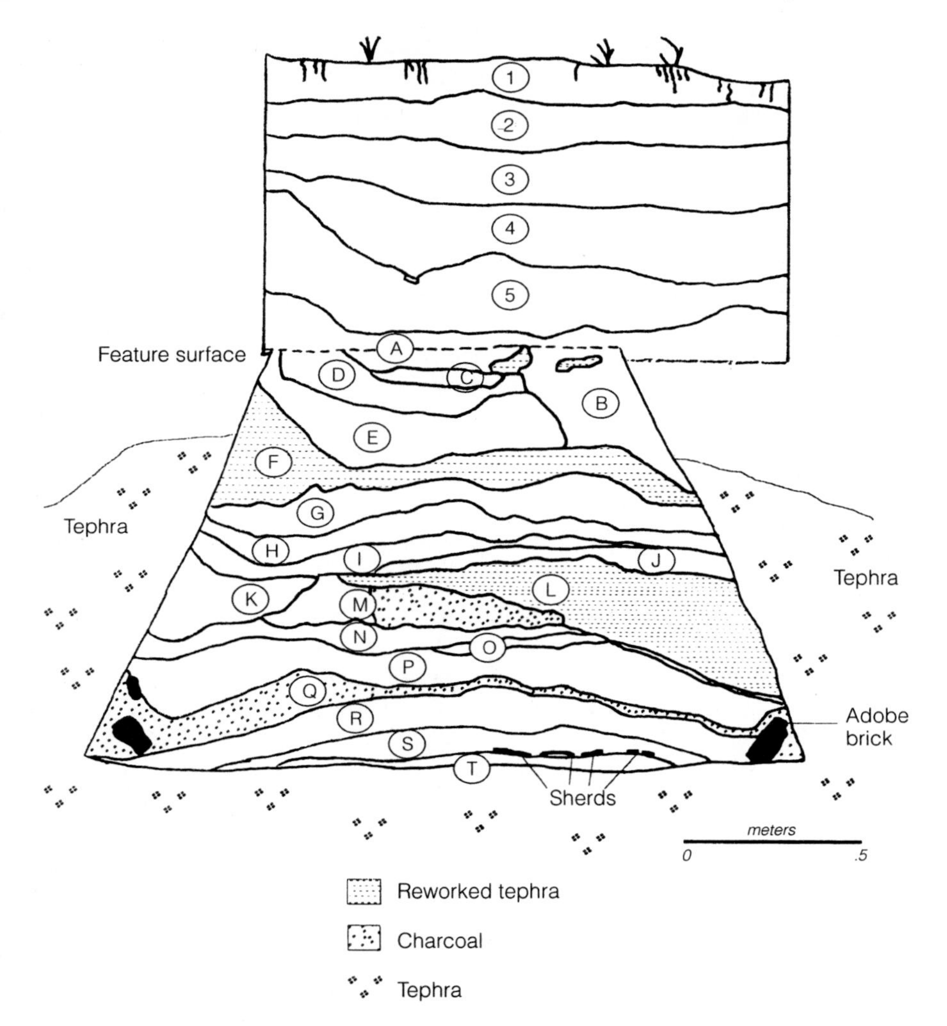

Figure 7.1 Profile drawing of Feature 5 (Big Pit) at the Pechichal site. Reproduced by permission of the Society for American Archaeology from Latin American Antiquity *11(3), 2000.*

These excavations are detailed in the 1994 project volume (Zeidler and Pearsall, 1994). At this time, most flotation samples taken from arbitrary excavation levels can only be affiliated to period, that is, to Jama Coaque I or II, rather than to Muchique phase. However, affiliation to Muchique phase is possible for most phytolith samples, since these were taken from stratigraphic deposits exposed in excavation profiles. As few charred macroremains were recovered from San Isidro flotation samples, phytolith samples in fact provide most of the available archaeobotanical data for the site. Formative period deposits are also present at San Isidro.

At Finca Cueva, located approximately 500 m west of San Isidro, six contiguous excavation units were opened and sampled for phytoliths and macroremains. Finca Cueva is a multicomponent site; excavations sampled in situ Muchique 1, Muchique

2, and Muchique 3 deposits, and also encountered mixed Formative period materials. Several features were excavated, but none are affiliated with Muchique 2.

The final site with confirmed Muchique 2 botanical samples is Dislabón, a multicomponent occupation site located west of the town of Eloy Alfaro. Two test units were excavated that encountered deposits affiliated to Muchique 1 and Muchique 2.

Table 7.1 summarizes the plants identified in Muchique 2–aged deposits. Since the richest plant assemblage comes from the Pechichal site, I will focus first on the patterning revealed in that data set, and then generalize to the nature of plant-people interrelationships during this phase for the valley as a whole (as represented by the five sites).

The discussion that follows relies not only on the *presence* of plant macroremains and phytoliths at sites, but also on macroremain *abundances*. As discussed in Chapter 1, macroremains are produced by the accidental or deliberate charring of plants, especially those cooked as foods. The more commonly a food is prepared, the more often it is exposed to fire and subject to accidental charring, or becomes part of burned refuse. Quantitative analysis can thus provide insight into the relative importance of different plant foods. A few definitions and notes on quantitative measures used in the rest of this chapter are therefore in order.

Density: charred ct/10 l (count of charred plant remains > 2 mm in size recovered per 10 liters of floated soil). This ratio is useful for assessing the overall abundance of macroremains recovered from a site. Very low densities may signal postdepositional destruction of macroremains, poor recovery technique, or lack of deposition of materials.
Richness: the total number of kinds of things present (in a sample, at a site). This is most often presented by the categories used in Chapter 6: numbers of kinds of economic plants present, for example.
Ubiquity (or more correctly, percentage presence): the number of proveniences (or samples) in which a taxon occurs (e.g., 50% maize presence = maize in half of all samples examined).
Frequency: percentage occurrence of a taxon, based on count (e.g., 10% *Trianthema* = 10% of the seeds in a sample/site are *Trianthema* by count). Usually presented as a pie chart.
Kernel:cupule ratio: the ratio of kernel to cupule fragments, based on count.
Maize:other food ratio: the ratio of all maize to all other food plants, based on count.

I sometimes also made a sum of all root/tuber macroremains (cf. *Manihot esculenta,* root/tuber, monocot root) and contrasted that count to the maize, arboreal, or bean counts. These comparison ratios are useful ways of showing change in an assemblage over time.

PECHICHAL RESULTS

Macroremains

The Big Pit, Feature 5 of the Pechichal site, is our best macroremain record of plant use in the Jama Valley, and provides a uniquely detailed view of early Jama-Coaque II subsistence (refer to Table 6.1, Table 7.1, and Appendix A). The rapidity with which this storage feature was filled with debris, the protective context, and

TABLE 7.1 PLANTS DOCUMENTED IN MUCHIQUE 2 PHASE (EARLY JAMA COAQUE II) CONTEXTS, EXCLUDING WOOD

Family Name	Identification	Common Name
Economic		
Cannaceae	*Canna*	*achira*
Euphorbiaceae	cf. *Manihot esculenta*	yuca
Fabaceae	*Phaseolus*	bean
Marantaceae	Marantaceae	arrowroot family
	Calathea	
	C. allouia	*llerén*
	Maranta	
	M. arundinacea	arrowroot
Malvaceae	*Gossypium*	cotton
Poaceae	*Zea mays*	maize
Cucurbitaceae	*Lagenaria siceraria*	gourd
	Cucurbita	squash
	Cucurbitaceae	gourd/squash family
Root/tuber		
Monocotyledon root		
Arboreal		
Annonaceae	Annonaceae	*guanábana* family
Arecaceae	*Aiphanes*	*coroso*
	Phytelephas	*cadi*
	Arecaceae	palm family
Bignoniaceae	cf. *Crescentia*	tree gourd
Bixaceae	*Bixa orellana*	*achiote*
Boraginaceae	*Cordia lutea*	*mullullu*
Chrysobalanaceae		
Dichapetalaceae	cf. *Tapura*	
Euphorbiaceae	*Croton*	
	Cleidion	
	Euphorbiaceae	spurge family
Hymenophyllaceae	*Trichomanes*	
Mimosaceae	Mimosaceae	tree legume
Myrtaceae	*Psidium*	*guava*
Sapotaceae	cf. *Sideroxylon*	dilly
	Sapotaceae	*zapote* family
Thick rind		
Thick flat rind		
Thick rind with internal divisions		
Thick layered rind		
Thin rind		
Thin rind with storied structure		
Curved rind		
Dense cotyledon fragment		
Small dense spherical fruit		
Dimpled fruit peel		
Thick rind with internal divisions		
Cystoliths		
Schlerids		
Unknowns: 82, 83, 87, 104		

(continued)

TABLE 7.1 (*continued*)

Family Name	Identification	Common Name
Open habitat/weedy		
Aizoaceae	*Trianthema portulacastrum*	
Amaranthaceae	*Amaranthus*	amaranth
	cf. *Iresine*	
Asteraceae	Asteraceae	sunflower family
Fabaceae	Fabaceae (744)	
	Fabaceae, beaked	
	Fabaceae, elongated	
	Fabaceae, small	
	Fabaceae, medium	
	Fabaceae, large	
Lamiaceae	*Teucrium*	
Linaceae	cf. *Linum*	
Malvaceae	cf. *Abutilon*	
	cf. *Herissantia*	
	Sida	
Musaceae	*Heliconia*	bird-of-paradise
Passifloraceae	*Passiflora*	passion fruit
Phytolaccaceae	Phytolaccaceae	pokeweed family
Poaceae	small-medium elongate	
	Panicoid	
	Festucoid	
	Chloridoid	
	Bambuseae/Arundinae	bamboo/cane
	Chusquea	bamboo
Portulacaceae	*Portulaca*	purslane
Rosaceae	cf. *Rubus*	raspberry
Solanaceae		nightshade family
Unknowns: 2, 6, 13, 18, 27, 29, 30, 32, 33, 34, 35, 51, 55, 59a, 85, 86, 91, 93, 97		
Other		
Porous endosperm fragments		
Cyperaceae		
Sponge spicules		
Diatoms		
Scitaminales		
Rugulose spheres		
Dicotyledon		
Epidermal appendages		

the relatively light overburden (excavated in Test Unit 1) all contributed to the excellent preservation of charred botanical remains. The cultural context of the deposits—disposal of (presumably) domestic debris after use as a storage pit ceased—is also an excellent one for looking at foodways.

Density of charred material, a measure of the quality of preservation of macroremains, was high throughout the Big Pit, as well as in Test Unit 1. Average count of charred material per 10 liters of floated sediment for the excavation unit was 135.9; for Feature 5, 148.5. Only three samples from the site had less than 50 pieces of charred material per 10 l. Element I, a relatively thin deposit in the middle of the pit, had the highest concentration of macroremains: 1,575/10 l. Density of macroremains was in fact high throughout most of the middle-lower section of the pit. The lack of any depth-related decline in density suggests that postdepositional destruction of charred botanical remains was minimal in the Big Pit.

Seven flotation samples were poppy-seed tested, resulting in an average recovery rate of 61% (see Chapter 4 for a discussion of this test). This is among the highest recovery rates for the project. Recovery of archaeological (ancient) small seeds is also the highest for the project, as I will discuss below.

Richness of macroremains (the numbers of kinds of plant taxa recovered) is very high in the Big Pit and the overlying excavation unit. All the economic plants identified during the project are represented. In the root/tuber category, monocot rhizome fragments and *Manihot esculenta* fragments are present (in two elements, each), as well as numerous root/tuber fragments. *Lagenaria siceraria* (bottle gourd) rind was recovered. Maize kernels and cupules are both abundant (1,762 kernel fragments; 2,340 cupules; kernel:cupule ratio = 0.75) and occur together in all but a few samples. Maize is ubiquitous in the levels of the test unit and feature elements. Cotton occurs in 68% of the proveniences, bean in 80%, and root/tuber in 64%. Gourd is less common, in 16% of proveniences. Of the 30 types of arboreal taxa that occur in Jama Valley samples, 70% occur in the Big Pit and Test Unit 1. Of the 60 or so types of small seeds present in flotation samples, 65% occur at Pechichal.

Looking at the occurrence of macroremains, excluding wood charcoal, in the Big Pit by count (Figure 7.2), maize is abundant, but arboreal taxa, root fragments, and open-area/weedy seeds are also common. Beans and cotton are represented at higher levels than seen at other sites. Thus while the ratio of maize:other foods is variable for individual elements (trash deposit episodes) in the pit, for the feature as a whole it is quite low: 1.1:1. This is a reflection of the broadness of the food base.

With a few exceptions, small seeds were not recovered in abundance from sites in the Jama River Valley. It is interesting, therefore, to look at what kinds of seeds occur in the Big Pit, and in what proportions (Figure 7.3). Note, for example, that while *Trianthema* and Solanaceae seeds are well represented (21%, 13%, respectively), these two commonly recovered seeds are only part of a much richer assemblage of taxa favoring open, disturbed habitats. Three types of Malvaceae seeds occur, for instance (*Sida, Herissantia, Abutilon*), as do six types of small wild beans or clovers (summed as Fabaceae on the pie chart; refer to Table 6.1). A number of these weedy taxa are useful. *Amaranthus,* Phytolaccaceae (pokeweed family), and *Portulaca,* among others, produce edible greens, for example, and *Rubus, Passiflora,* and some types of Solanaceae (i.e., ground-cherry) produce edible fruits, as discussed in Chapter 6.

I have not discussed patterns of wood use at sites investigated during the Jama project because of difficulties in identifying wood mentioned earlier (refer to Chapter 6). I was interested, however, in how richness of wood taxa present at Pechichal might compare to other Muchique 2 sites, given the excellence of preservation of macroremains at Pechichal. We identified wood from nine flotation samples from the Big Pit. Twenty-seven wood morpho-types were tallied in these samples. The most commonly occurring types, those present in more than half of the samples, were cf. *Platypodium* (Fabaceae, Unk 6), cf. E1663, *lengua de vaca* (Unk 10), cf. *Aiphanes* (Arecaceae, Unk 13), and cf. *Mimosa* (Fabaceae, Unk 20). By contrast, only 13 wood morpho-types occurred in five samples analyzed from the Feature 4 floor at the El Tape site. At El Tape the most commonly occurring wood types were cf. *Platypodium* (Fabaceae, Unk 6), cf. E1663, *lengua de vaca* (Unk 10),

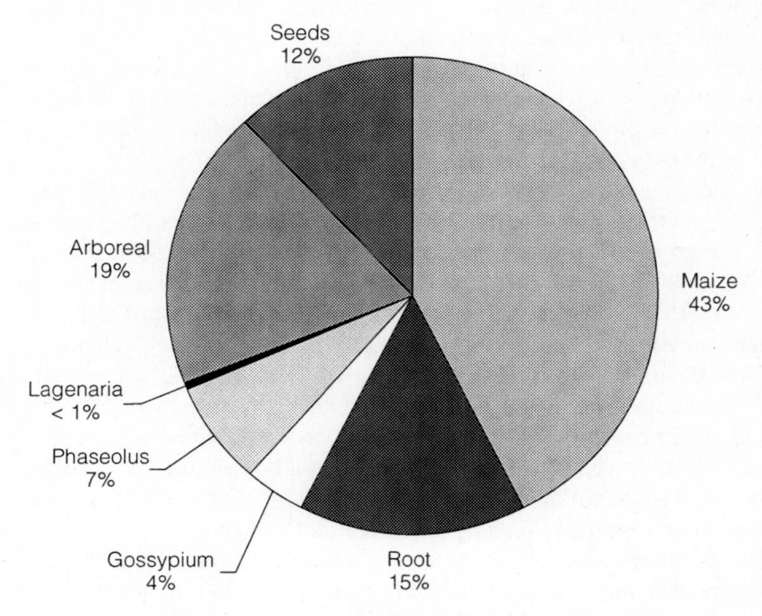

Figure 7.2 Percentages of types of macroremains recovered from the Big Pit

cf. Fabaceae (Unk 22), and Unk 31. Wood was abundant in the samples from both sites; more than just dense woods were preserved.

These findings suggest, then, that the Pechichal wood assemblage is richer than the El Tape assemblage, with double the number of types of wood present. The sites are similar, however, in exhibiting near ubiquity of cf. *Platypodium,* a type of tree legume. Both sites are located in the Jama bottomlands, and differences in forest composition between the middle and lower valley are not dramatic, although the lower valley is drier. Cultural, rather than environmental, factors may explain the differences in wood selection, but it is difficult to explore this issue with the limited database available to us at this time.

Phytoliths

Each element in the Big Pit was sampled for phytolith analysis. Six of these samples were selected for analysis: Elements C, F, I, M, O, and R. The results are presented in Figure 6.3.

In an alluvial setting like the Pechichal site, phytoliths are deposited not only by human agency, but by fluvial action. Phytolith assemblages are thus a combination of in-situ decay of residues of plants brought to the site, used, and disposed of, and phytoliths from plants growing in the watershed, and deposited through the action of the river. Phytoliths may also be removed from fluvial sediments by erosion.

It can be difficult to distinguish between phytoliths deposited through natural and human agency. In an alluvial setting, however, the analyst must assume that all samples contain numerous phytoliths that represent the "background" vegetation of the

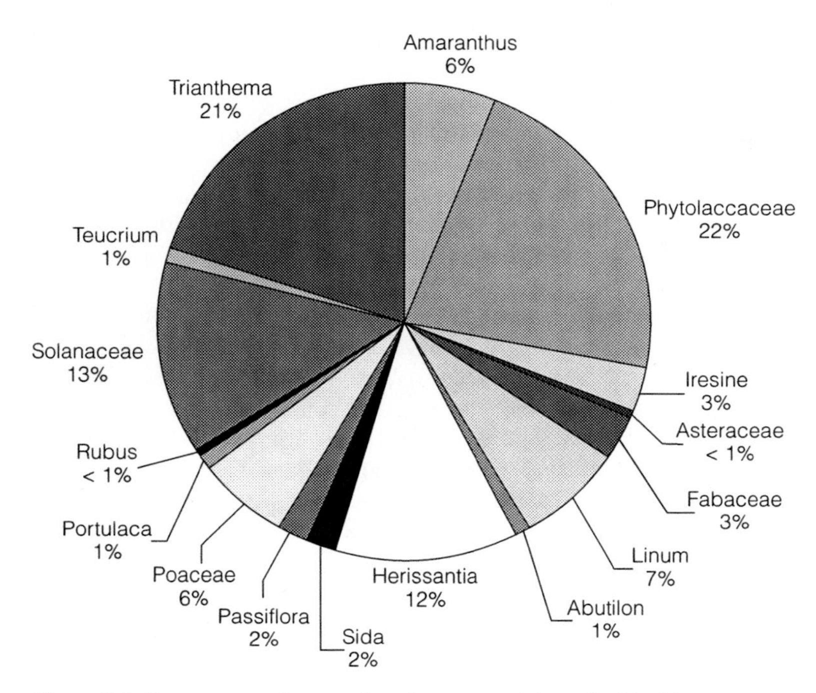

Figure 7.3 Percentages of types of seeds recovered from the Big Pit

watershed. Notice in the Big Pit samples how the proportion of open habitat, arboreal indicators, and wet habitat indicators is quite similar among all elements (except for Element O). This in all likelihood reflects the local environment (moist alluvial soils) and the balance of trees and cleared areas in the watershed. Considerable forest clearance has taken place by this time, as I'll discuss further in Chapter 9. In the case of the Big Pit, however, I suspect that the contribution of phytoliths from decaying/burnt food refuse is in fact quite high. In several elements, over 65% of phytoliths are open-area indicators. Given the abundance of charred seeds of open-area plants, this likely represents the "background" level plus the in-situ decay of plants harvested for edible leaves, seeds, and fruits. Similarly, the 35% occurrence of Chrysobalanaceae phytoliths in Element O most likely represents deposition from decay of residues of edible fruits in this family (e.g., *Chrysobolanus icaco*), rather than a resurgence of forest.

Other useful plants documented in Pechichal phytolith samples include maize (maize leaf and cob bodies are both present), *Canna, Maranta, M. arundinacea, Calathea, C. allouia,* and Marantaceae. The *Canna* spheres present in samples are likely from cultivated/escaped *Canna tuerckheimii* (= *C. edulis*). As discussed earlier, *Calathea allouia, llerén,* and *Maranta arundinacea,* arrowroot, are cultivated plants in the Marantaceae, and there are also wild species present in the valley. Phytoliths from palm and Annonaceae (*guayábana* family) also occur.

In summary, the excellent preservation of charred materials and the abundance of phytoliths recovered in Feature 5 of the Pechichal site give us a glimpse of the richness of the food base of the Muchique 2 phase population occupying this site, a food base that included both cultivated and wild/tended food resources, and root and seed crops. The fact that fragile charred root/tuber fragments are very common at the site suggests that foods such as yuca (*Manihot esculenta*), *achira* (*Canna*), *llerén* (*Calathea allouia*), and arrowroot (*Maranta arundinacea*) were important components of diet: such remains tend to be underrepresented relative to sturdy materials such as maize cupules and tree fruit fragments. Further, the fact that charred remains and phytoliths of root/tuber crops were found in direct association with maize (i.e., in the same trash deposits) suggests that root/tuber foods and maize were both part of daily, domestic cuisine.

Foods were not limited to cultivated plants at the Pechichal site, however. Fruits of palm, sapote family, tree legumes, and guava, among other arboreal taxa, were commonly utilized. A diverse array of seeds of weedy plants were also recovered. *Amaranthus,* Phytolaccaceae (pokeweed family), and *Portulaca,* among others, produce edible greens, and *Rubus, Passiflora,* and some types of Solanaceae (i.e., ground-cherry) produce edible fruits. Forest clearance—most likely for agriculture, but also a by-product of volcanic ash (tephra) fall—created favorable environments for weedy growth, and some of these plants were likely used as greens, fruits, flavorings, or for medicinal purposes.

The common occurrence of fragile cotton seeds in the Big Pit suggests that this was an important industrial crop for the inhabitants of the site. *Achiote (Bixa orellana),* a useful dye plant, was also recovered.

RESULTS FROM OTHER SITES WITH MUCHIQUE 2 DEPOSITS

The El Tape site provides a Muchique 2 sample from the lower valley (Stratum I). Feature 4, an early Muchique 2 floor, was extensively sampled for macroremains, and phytolith samples were taken from this feature, as well as a second feature and a deposit affiliated with Muchique 2. Charred remains were very dense in Feature 4: 518 pieces of charred material per 10 liters of soil. Most recovered macroremains were wood charcoal, but food remains also occurred in some abundance. In fact, charred residues of cooking activities made up a considerable part of the matrix of this Muchique 2 floor.

While the density of charred remains was high at El Tape, richness (the numbers of kinds of botanical remains recovered) was relatively low in Feature 4. Of the economic plants recovered at Jama Valley sites, maize, root/tuber, and beans occur as macroremains in Feature 4, and Cucurbitaceae, *Canna,* and Marantaceae phytoliths were all recovered. However, of the macroremains, only maize was abundant. Of the tree fruits recorded in Jama Valley samples, only 6 occur as macroremains in Feature 4 (thick rind, thick rind with internal partitions, thin rind, curved rind, dense cotyledon, and small dense spherical), and palm, Annonaceae, and Chrysobalanaceae phytoliths are present. Of the 60 or so kinds of charred seeds that occur in Jama samples, only 1 occurs in Feature 4 (Unk 6). *Heliconia* and grass phytoliths are also present. Especially interesting is the absence of the most commonly recovered seed types, *Trianthema portulacastrum* and Solanaceae. It seems, then, that the refuse recovered from the early Muchique 2 floor (Feature 4) at El Tape documents food preparation activities focused on maize, and that the activities that resulted in the charring of maize kernels and a few other remains did not incorporate weedy seeds into the macroremain assemblage.

The maize focus of the Feature 4 assemblage is clearly illustrated by the fact that charred maize remains make up 92% of the Feature 4 assemblage, by count. The maize:other food ratio (14:1) is another reflection of the dominance of maize in the Feature 4 assemblage. Finally, it is interesting that the vast majority of maize remains recovered from Feature 4 were kernel fragments; very few cupules were preserved. This suggests that shelled maize was being utilized at El Tape.

San Isidro and Finca Cueva are located on the same alluvium in Stratum III. Twelve Muchique 2–affiliated flotation samples are available from Finca Cueva; no phytolith samples are available from these contexts. Density of macroremains in Muchique 2 samples at Finca Cueva ranged from 1.3/10 l to 12.9/10 l, with one sample with a higher density (62.3/10 l). Density drops off at around 60 cm below the modern surface (Muchique 3/Muchique 2 boundary in the deposits). In all likelihood, the low density of Muchique 2 macroremains is partially a depth-dependent phenomenon, with charred materials broken up by the weight of overlying soil, and repeated wetting and drying cycles of the alluvial deposits.

As might be expected, the low density of macroremains in Finca Cueva deposits results in relatively low richness. Of economic taxa, only maize and root/tuber are present. There are six types of arboreal plants, palm, tree gourd, dilly, and three morpho-types. Small seed taxa include nine types, cf. *Iresine,* cf. *Linum,* cf. *Herissantia,* Poaceae, Solanaceae, *Trianthema portulacastrum,* and three unknowns. By count, small seeds dominate the assemblage (80%), with arboreal taxa making up most of the rest (17%). Maize makes up 3% of the assemblage by count, but occurs in 75% of the samples. The fact that maize is widely scattered, but low in count, again suggests a degraded assemblage: macroremains lost due to postdepositional destruction.

While no Muchique 2–affiliated phytolith samples are available from Finca Cueva, deposits and features affiliated with Muchique 1 document that *Canna,* Cucurbitaceae, and Marantaceae were also present at this site during Jama-Coaque times.

As I mentioned earlier, the archaeobotanical record from San Isidro, the regional ceremonial center of the middle Jama Valley, is not as strong as we would like. This is due in large part to the difficulties of excavating at a complex site with deeply buried deposits under a bustling, modern town, and in part to small flotation sample size used early in the project. Further, work is still in progress to assign Jama-Coaque II deposits at the site to phase (i.e., to either Muchique 2 or 3).

Phytolith samples from several deep profiles at San Isidro do yield insights into plant use at the site during Muchique 2 times, as well as into the nature of vegetation in the surrounding area. Open habitat indicators, especially grasses, dominate phytolith assemblages at the site; Asteraceae and *Heliconia* also occur. Among useful arboreal taxa documented are Annonaceae, Chrysobalanaceae, and Arecaceae (palm). Maize, Marantaceae, and *Canna* also occur.

Two flotation samples affiliated with Muchique 2 are available from the Dislabón site, located in the upper part of Stratum III. Density of macroremains is fairly low in these samples (average: 13.7/10 l). The macroremain assemblage is not very rich; recovery included only one economic plant (maize), five types of arboreal plants (including palm and guava), and two types of small, charred seeds. By count, small seeds are the most abundant macroremain (47%), followed by maize (30%) and arboreal taxa (23%). Phytolith analysis of one Muchique 2–affiliated deposit adds Chrysobalanaceae, *Canna,* and Marantaceae to the list of useful plants. However, this sample lies just 8 cm below the modern surface, and the presence of phytoliths from banana, a crop introduced during colonial times and widely cultivated today on alluvial soils, suggests that this sample is mixed with modern sediments.

PLANT-PEOPLE INTERRELATIONSHIPS DURING MUCHIQUE 2

Study of macroremains and phytoliths from Jama Valley sites with deposits affiliated with Muchique 2 reveals a subsistence base that included domesticated, tended, and wild plant species; root/tuber crops and seed crops; and annual and perennial resources. The richness of utilized plants in well-preserved assemblages reveals the great diversity of plant resources in this dry tropical forest setting. Maize and beans (annual seed crops) and yuca, *achira, llerén,* and arrowroot (perennial root/tuber crops) are among the domesticated food crops grown in the valley. Gourd and squash were also grown. Fruits of at least two types of palm, *coroso* and *cadi,* and undoubtedly more, were probably eaten, as were fruits of *Psidium,* Annonaceae, *Sideroxylon,* other Sapotaceae, and Chrysobalanaceae. In addition to these identified arboreal resources, many other fragments of tree fruits were recovered that could not be identified. These add to the richness of the subsistence base provided by wild or tended perennial species. This combination of utilized resources suggests that a mosaic of open and forested habitats, with abundant edges, existed in the valley by Muchique 2 times, an interpretation supported by faunal data from the Big Pit and phytolith analysis of the Rió Grande profile (see discussion in Chapter 9).

None of the small seeds documented in the Muchique 2 record appear to be significant food sources. A few are useful plants (there are some potential medicinals, and a few edible greens or fruits, as discussed in Chapter 6), but there is no evidence of a starchy or oily "small seed complex" that played a significant role in subsistence. I have collected nearly all the identified taxa along roadsides, in overgrown agricultural fields, and similar disturbed habitats. These taxa thus may be incidental inclusions in the record that reflect the open habitats in and around sites. Having said this, greens and seasonal fruits of plants taking advantage of open and edge habitats probably provided variety, vitamins, and minerals to the diet, and served as a source of medicinal or ritual plants.

In addition to food plants, a number of plants identified in Muchique 2 samples provided materials for construction, household implements, clothing, and containers. Gourds and tree gourds are both useful as containers, serving utensils, or net floats. Bamboo and wood provide numerous uses, including wall construction and household objects such as sleeping platforms. Palm, grasses, and bird-of-paradise are all useful for roof thatch. Cotton was likely spun and woven into cloth, or netted to make hammocks, carrying nets, and fishing nets. *Achiote* provided a red dye for cloth, body paint, or food coloring.

It is difficult to compare results from sites with excellent macroremain preservation and recovery (Pechichal, El Tape) to those with degraded assemblages (San

Isidro, Finca Cueva, Dislabón). With the exception of San Isidro, very few cultural contexts at each site have been investigated. For these reasons, I have mostly focused on patterns revealed by the Muchique 2 data set as a whole, rather than trying to look at differences and similarities among sites. It is interesting to note, however, what may be specialization in crop production at Pechichal and El Tape. At Pechichal, both maize cupules and kernels are preserved, suggesting that maize was both grown and processed/consumed at the site. At El Tape, by contrast, few cupules were recovered, while kernel fragments were abundant. This may be an indication that shelled maize, rather than maize still on the cob, was processed. Shelled, dry maize is more efficiently transported and stored than maize on the cob. As discussed in Chapter 6, it appears that the same type of maize was used at both sites. The Pechichal pit also preserved abundant cotton seeds, as well as the dye plant achiote. Production of cotton and cotton cloth may have been a specialization of this site.

How and where were the crops documented in the Muchique 2 archaeobotanical record grown? What was the impact of agricultural activities and tephra fall on the environment, and how did Muchique 2 populations respond to, and survive, an ash fall event more severe than those that had driven earlier populations from the valley? These issues are addressed in the next two chapters.

8/Producing Food in the Jama River Valley

A useful starting point for understanding how agriculture may have been practiced in prehistory in the Jama River Valley is understanding the nature of contemporary traditional agriculture. As I discussed in Chapter 3, the Jama River Valley, with much of Manabí Province, was largely depopulated early in the Colonial period. The late 19th and early 20th centuries saw the resettlement of the Jama region by people who had no cultural link to the valley's prehistoric populations. However, these settlers and their descendants faced challenges of making a living in this dry tropical forest setting similar to those faced by the Jama-Coaque people. What are these challenges, and how can they be overcome? How productive is the valley for growing maize (*Zea mays*) and yuca (*Manihot esculenta*)?

To begin to answer these questions, I present an overview of maize and yuca cultivation today, compiled from my field observations in the valley. This is followed by a detailed analysis of maize and yuca yield data, and an evaluation of several hypotheses about productivity in different zones of the valley. I then present a short discussion of the results of my survey of contemporary land-use patterns. Finally, I develop a working model for the reconstructed agricultural landscape.

MAIZE AND YUCA CULTIVATION TODAY

Three varieties of maize are commonly grown in the Jama River Valley today: *amarillo (criollo), cubano blanco,* and *hibrido* (Figure 5.4). *Amarillo* is the local name for *candela,* a traditional lemon-yellow to bright orange maize with floury kernels. Timothy and colleagues (1963) note that this variety resembles the flour corns of the Amazon basin. *Cubano blanco* is perhaps a recent introduction of maize common in the Caribbean, but is considered a "traditional" variety by local farmers. Kernels vary from flint to dent in texture, and from white to yellow, to light orange in color. *Hibrido* (hybrid, or improved maize), specifically *hibrido* 515 and *hibrido* 526, makes up the third class of maize commonly grown. These improved varieties are usually grown from saved seed, and grade into *cubano blanco* in appearance. A few farmers still grow popcorn (*canguil*).

Farmers select the variety of maize to grow in part based on their expectations of how the varieties perform under different field conditions. *Hibrido* 515 is grown, for example, on less fertile ground—that is, when the ground is "tired" from three or four years of continuous maize cultivation. *Hibrido* 515 is also considered good for hilly ground, since it is short. Of the traditional varieties, *cubano blanco* does better on tired ground than does *amarillo. Amarillo* does best in flat, humid ground, or on fertile land that lies at higher elevation, where it is moister. In general, the traditional races are more likely to be grown up in the hills than the hybrid varieties. Neither *cubano blanco* nor *amarillo* perform well in a drought, however. I documented one case in which a farmer planted *amarillo* and *hibrido* 515, and there was little rain. None of the *amarillo* plants produced ears, but the farmer harvested 15 *fenegas* of *hibrido* 515 from a total of 30 *cuadras* planted in the two varieties. (See Chapter 5 for descriptions of these units: a cuadra is about 0.7 ha; a fenega is 1,800 ears of maize.)

Amarillo produces *choclos* (green maize eaten like sweet corn) in about 65–75 days and well-dried *mazorcas,* or mature ears, in 6 months. It is primarily grown for *mazorcas,* and the flour is prized for making traditional foods. The hybrids develop faster than the traditional races, producing *choclos* in less than 65 days, making them the variety of choice for farmers growing green maize for sale.

The most common yuca variety grown in the valley today is *yuca de tres meses* (3-month yuca), sometimes referred to as *yuca morada* or *yuca roja* because of the reddish skin of the roots. *Tres meses* is preferred by farmers who want a quick yield, for example, as a single harvest for starch production or for livestock feed. Other farmers prefer yuca that matures more slowly and does not have to be harvested when other crops are maturing. *Tres meses* can in fact be left in the ground for up to 18 months. *Espada blanca,* also called *yuca de seis meses* (6-month yuca), is also present in the valley, and a few farmers grow *yema de huevo* or *yuca negra,* two varieties that require a year to mature. *Seis meses* can be left in the ground about two years, and *yuca negra* is said not to rot in the ground for up to three years. A farmer may leave yuca in the field until he or she needs the money from selling it. All yuca varieties grown in the Jama Valley are "sweet" manioc (low in cyanic acid) and do not require any processing before cooking except removal of the thick, woody peel.

The agricultural cycle for maize and yuca cultivation starts in the late dry season, in October or November, with the "cleaning" of last years' fields or clearing of new fields from forest, abandoned CCP (café, cacao, plátano) plantations, fallow fields, or pasture. Abandoned or fallow (resting) fields quickly regrow into a tangled secondary forest (Figure 8.1). These can be cleared and reseeded in maize after three years. Cleaning of fields is timed so that the fields are ready for planting at the beginning of the rainy season. In a "typical" year, planting takes place in late December or early January. Farmers usually wait to plant maize until the rains have actually started, when the soil is already moist and it looks like rain will be steady. Planting too early can result in loss of newly sprouted maize plants if the soil dries out. Yuca may be planted earlier, as soon as fields are prepared.

Preparation of new fields in high forest involves felling trees and clearing underbrush. Some farmers burn off the fields; others do not. In the case of an incomplete burn, crops are planted among the fallen trunks (Figure 8.2). If the field is not burned, weeds and residues of last years' crops—if present—will at least be gathered into piles between the rows of crops. Farmers report more problems with weed growth if fields are not burned, but early onset of rain can result in poor burns. The

Figure 8.1 Clearing fields for agriculture: a. typical tangled growth of a field allowed to regrow into secondary forest

work of preparing a field is often done in stages: the farmer may cut the trees in a "new" field first, machete weeds and crop residues from adjoining fields planted in prior years later, then burn both areas after brush and trees dry for a few weeks.

Cutting trees or brush, cleaning up weeds and crop residues, and stacking and/or burning requires from 3–12 *jornales* (man-days), with an average of 7 jornales/cuadra, for maize plantings. This work is the labor of men and boys; farmers usually hire help to supplement the work of family members. The primary tool used is the steel machete. A day's work includes the midday meal. I have less data on field preparation for yuca; two estimates are 12 and 18 jornales for preparing a one-cuadra field. Often maize and yuca are intercropped, as described below. Field size varies, but 1–4 cuadras is a typical range for maize or yuca plantings located away from the house. Most families maintain a small garden adjacent to their home, and this often contains a few yuca plants mixed in with fruit trees, herbs and spices, ornamentals, and medicinal plants (Figure 8.3). Kernels from the center of the maize cob are used for planting, and cobs with straight rows are selected. Maize is planted by men, each walking in a line along separate rows. A heavy pole is dropped to make a planting hole, into which 3–4 maize kernels are placed. This planting group is called a *mata*. Fewer seeds of *cubano blanco* are planted per mata, usually 3, and more seeds of *amarillo,* usually 5. Planting depth varies; some farmers prefer to

Figure 8.1 (continued) b. secondary growth cut and ready to burn

Figure 8.2 Planting a recently burned-off field. In this case the burn was incomplete. Notice the long poles (planting sticks) in the workers' hands.

Figure 8.3 This house garden is small, but contains a rich array of ornamental and useful plants.

plant deep (15 cm or so) so birds will not steal the seeds, while others plant just under the surface. Spacing between matas and between rows varies, as discussed below. Spacing is measured in *varas,* with typical spacings of 1, 1.5, or 2 varas between matas and rows (one vara is 0.84 m, a stride more or less). Farmers report that it takes from 2 to 4 jornales to plant a cuadra of maize (average, 3.5 jornales).

Yuca cultivation is less tied to a seasonal schedule than maize. For household use, yuca is harvested as needed: the roots store well in the ground, but rot rapidly after harvest. After a mata is pulled up and the roots broken off, a farmer will often cut off a piece or two of stem and replant immediately. (Note the leaf nodes shown in Figure 5.5; these will sprout when the stem is planted.) As long as there is enough moisture in the soil, the stem cutting will put out roots and the young plant will become established. It will grow during the rainy season, and go dormant, dropping its leaves, when the rains taper off in April. So although it is common to see farmers planting yuca in a new field just before or after maize is planted, established fields may contain yuca of various ages. The exception to the plant-by-plant harvest pattern is harvest for starch production (see below) or for sale. In this case, the farmer may replant the entire field right after harvest, that is, at the end of the rainy season, perhaps putting cuttings between drying maize plants. This allows the yuca to take advantage of late rains.

As mentioned above, yuca is planted by stem cuttings. On a slope the stem is usually inserted at an angle, with the slope, leaving about half of the stem out of the ground. The angle makes it easier to pull the roots out at harvest. Another way to plant is to put the stem into the ground at an angle, but completely under ground. This delays sprouting somewhat, allowing work to continue in the field without damaging the plants. Some farmers plant short stem pieces, *una cuarta* (a handsbreadth), while others prefer longer cuttings (50 cm). One farmer explained that he preferred to plant long stems upright, rather than with the slope, so that the yuca grew above the quickly sprouting weeds rather than being buried by them. In reality, how yuca is planted seems to be a matter of personal preference; yuca is a very tolerant plant. Two to four jornales are needed to plant a cuadra of yuca.

When maize is planted between rows of yuca in an established field, the yuca plants may be cut back at the first maize weeding, when the maize plants are about knee-high, to ensure that the young maize is not shaded. By the time the yuca leafs out again, the maize is well established. Farmers will weed at least once more, when the maize is about chest high, but usually do not cut the yuca back again. I was told that by cutting yuca plants back once, they grow well for a longer period. *Tres meses,* for example, can be grown into a second rainy season, or for up to 18 months. Anticipated yield from such old yuca plants is 25–30 lbs of roots/plant. The intercropping pattern mentioned above, that is, alternating rows of maize and yuca, is the most common one that I observed. Less common is mixing maize and yuca in the same row, or planting adjoining blocks of pure stands of each crop.

How long yuca takes to mature—to produce good, heavy roots—depends on the rain. If yuca is planted in January but the rains are poor, farmers will let most plants sit in the ground until the following rainy season, when two months of good rain will finish them up. Only 10–12 lbs/plant are expected if yuca is harvested during its first growing season. Some farmers wait until the dry season to harvest yuca because the roots are "softer" then: the starchy roots serve as a food reserve for the dormant plant.

Yuca can be grown on the same piece of land for about four years. After this, yields fall off, and most farmers will rest the field for two to three years, then replant it. It is not necessary to burn a fallow yuca field if it is not heavily overgrown; cutting the weeds is sufficient. Upland soils are considered good for yuca, but land nearer the river is better, since it is sandier and moister. Farmers may grow yuca among café, cacao, and plátano plantings until the CCP plants start to shade the ground too much. This practice allows the yuca to take advantage of any irrigation for CCP plantings, and to benefit from the rich soil. A few farmers still use runoff irrigation in fields—that is, they build simple catch dams and dig canals to guide runoff water from rains into fields planted at the bottom of slopes. The presence of abandoned dams and canals in some fields suggest that runoff irrigation was once more widespread, but we do not have systematic data on these features.

Some yuca is processed locally for starch extraction (Figure 1.1d). Processing may be done by someone other than the owner of the yuca, in which case the owner and the processor divide the starch between them, or people will get together and process quantities of starch. Peeled and grated roots are put in a cloth with some water. The cloth is squeezed or wrung and the liquid caught in a container. After the starch settles out, the liquid is poured off and the starch is allowed to dry. The fresh mash, after being processed for starch, can be used to make tortillas, but it tastes better with some of the starch remaining. The starch is used like flour or sold.

Most farmers told me that maize can be planted in the same field for three years, although a few said four or five years. Only in *terreno bajo,* alluvial land near the river, can maize be grown every year. When harvests decline, land is planted in *higerilla,* an oil crop, or, more commonly, seeded with pasture grass. Although one farmer told me of taking land out of pasture after four to five years and planting maize again, this is not typical: increasingly pasture is the "climax vegetation" of the Jama Valley (refer to Figure 3.4). It is more common to cut forest for new maize fields after putting old fields into pasture. These new fields often adjoin older planting areas.

The maize harvest starts in mid-March with the harvest of *choclos*—green maize ears. *Choclos* are harvested gradually over about eight weeks, until kernels get too hard, or cut all at once for sale. By June maize is fully mature and dry (the *mazorca* stage), and fields are being actively harvested all over the valley. If there are late rains, however, farmers leave the ears in the field to dry for a longer period. Weeds are cut in fields just prior to harvest to drive away snakes and make access easier. Harvesting is done by hand.

Rain in March is critical for the development of large ears with good kernels. Drought late in the growing season can result in ears that are all cob and few kernels. Maize may be half the size in a dry year as in a wet year, and much will not finish ripening. Farmers throughout the valley agree that the success of maize depends on the rains, and only secondarily on soil quality. Café, cacao, and plátano plantings, by contrast, must be grown on fertile "black" soil. Final maize ear size is also affected by planting distances. If there is good rain, then closely planted maize does well. But if it is dry, then closely planted matas produce smaller ears than matas spaced further apart. In general, the further apart the matas are planted, the larger the ears will be. Thus farmers believe there is a trade-off between increasing density of planting (closer spacing) to increase numbers of ears and decreasing the density (wider spacing) for larger ear development (refer to Figure 5.6).

What is a "good" maize harvest in the Jama Valley? There are differences of opinion about this that relate to individual farmers' experiences and land access, but in general 6–7 fenegas/cuadra is considered a very good yield, 4–5 fenegas/cuadra a good yield, and 3 fenegas/cuadra is described by most as *regular* (okay, or average). Less than 3 fenegas/cuadra is a poor yield. (Recall that a fenega is 1,800 ears.) It is my impression that poorly developed ears, those with poor kernel set, are not counted in the harvest. Since relatively little alluvial land is devoted to maize production, these yields most likely characterize production in upland fields.

In this chapter I focus my discussion on maize and yuca cultivation because these are the traditional crops that are still widely grown in the valley, and farmers devote most of limited land available to them for rainy season plantings to these crops. It is rare to walk through a field and not see a variety of other useful plants tucked in among the maize and yuca, however. Even on the steepest, upland fields, farmers will plant a few matas of squash or watermelon and hope for a few fruits. If a family has access to alluvial lands, pole bean, bean, ají (chile pepper), watermelon, and sweet potato, among other annual crops, will be interplanted with maize and yuca in the mixed plantings typical of tropical forest agriculture (Figure 8.4). Useful tree species are left on the edges of fields and harvested while the field is in use and after it is fallowed or planted in pasture. Palms are among the most common trees left in cleared fields, since palm thatch is still prized as roofing material for traditional houses.

ANALYZING BASELINE DATA ON MAIZE AND YUCA PRODUCTIVITY

I now present results of the study of modern maize and yuca yields in the Jama Valley, and interpret those results in terms of the potential productivity of the valley for sustaining its prehistoric inhabitants. In Chapter 5 I described the procedures followed in this study. It is important to stress again that my goal in conducting the agri-

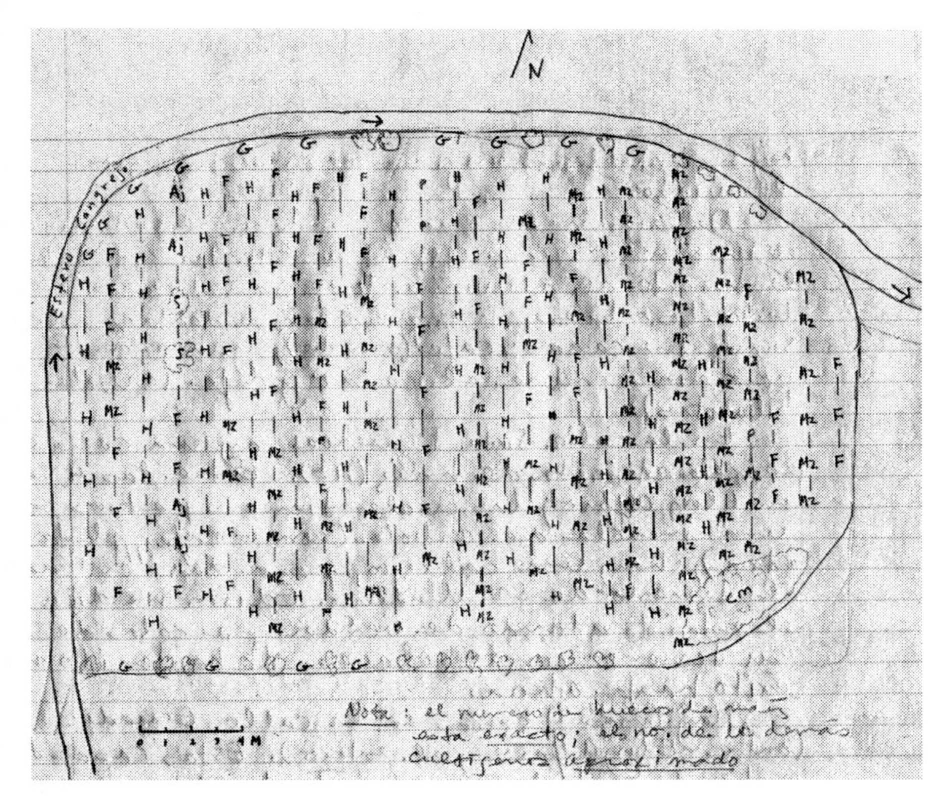

Figure 8.4 Intensely planted field in the alluvium of the Estero Congrejo in San Isidro. Crops include pole beans (H), maize (mz), common beans (F), watermelon (S), chili peppers (aj), and banana (G)

cultural research—gathering the yield data and talking to farmers about agricultural practices—was to model the *potential productivity* of the valley. There is no way we can know how far apart maize plants were spaced by Jama-Coaque II farmers, or what type of interplanting, if any, they preferred. We do not have buried fields with plants preserved in situ. We do not know how many fanegas/cuadra of maize were produced during any period in the past. What we can discover, however, is something about the variability of maize and yuca productivity around this valley when traditional cultivation practices are followed (i.e., in the absence of fertilizer, herbicides, systematic irrigation, and mechanized cultivation). This gives us upper and lower limits to use in modeling the ways in which people could have successfully adapted to this setting in the past, and tests some of the expectations and generalizations of modern farmers.

As discussed in Chapter 5, several hypotheses structured the way I collected the agricultural data, and my subsequent analysis of it. Formally stated, these hypotheses are:

H1. That alluvial lands are more productive for maize and yuca agriculture than nonalluvial lands, and that productivity also differs between major orders of nonalluvial soils

H2. That agricultural productivity in the uplands (nonalluvial soils) increases as elevation increases

H3. That there are differences in agricultural productivity among Stratum I, Stratum II, and Stratum III that might explain the early intensive settlement of the San Isidro area in Stratum III

After discussing the soils of the valley, I examine each of these hypotheses in turn, using maize and yuca yield data, and sketch a preliminary model for past agricultural productivity in the valley.

Agricultural Soils of the Jama River Valley

Crop productivity is dependent on a number of factors, chief among them the abundance and timing of rainfall and soil fertility. I classified soils of fields studied during the project using a soil map of the region and consulted several soil texts (Buol et al., 1997; Landon, 1991; van Wambeke, 1992) to learn about the properties of tropical soils. To classify the soils, I overlay the soil map on the topographic maps showing field locations, and classified each field by soil type. Fields studied for maize or yuca yields were located on 15 different soils (Table 8.1).

Soil taxonomy is complex, and valley soils fall into a number of soil orders, suborders, and great groups, as shown on the table. The three fluvially deposited soils are all deep soils of variable texture, renewed by deposition of new soil, and so favored for traditional agriculture. The most common coastal alluvial soils (28 in Table 8.1) are more alkaline than the inland alluvium (27) and have vertisol-like features (a tendency to swell when wet and shrink when dry) that are lacking in the inland soils. Most of the alluvial soils in the Jama Valley are young and have little to minimal horizon development.

The other 12 types of agricultural soils are not fluvially deposited. Eight of these are mollisols: well-drained forested or grassland soils that are relatively deep and fertile, and characterized by well-developed topsoil. Drier mollisols (ustolls) dominate Stratrum I; moister mollisols (udolls) occur in the western part of Stratrum II and in southern and southwestern parts of Stratum III. Some of the coastal mollisols have vertisol-like features.

Four of the nonfluvial soils are inceptisols: young soils that show some profile development, but few other diagnostic features. Most inceptisols in the Jama Valley have vertisol-like features. In general, inceptisols are mineral soils with medium to fine texture. Many form on steeply sloping terrain. This is the case in the Jama Valley; most of Stratum II is classified as alfisol and inceptisol, as is the northeastern part of Stratum III.

After the 1988 season of the agricultural yield study, I submitted soil samples collected in fields during 1982 and 1988 to the University of Missouri Soils Laboratory for analysis. Dr. Randy Miles of the MU Soils and Atmospheric Science Department examined the test results for me. The following is a summary of his assessment of fertility in the sampled fields.

The overall impression of the soil analyses is that soil nutrients and organic matter are favorable for plant growth, and that there are more similarities than differences among samples tested from fluvial and nonfluvial fields, and nonfluvial fields on different soils. This finding should perhaps not be surprising: the fields I sampled

TABLE 8.1 SOILS OF AGRICULTURAL PLOTS STUDIED DURING THE JAMA PROJECT

<table>
<tr><th>Code No.</th><th>Great Group Classification</th><th>Soil Order</th></tr>
<tr><td colspan="3">I. Coastal mountain range</td></tr>
<tr><td>1</td><td>Tropudalf and Eutropept</td><td>Alfisol and Inceptisol</td></tr>
<tr><td colspan="3">Clayey soils of variable depth, locally colluvial. Rocks, debris, and rocky outcroppings present. pH 5, 5–7. Tropudalfs are continually warm udalfs, which are moderately wet alfisols. Eutropepts are tropepts, tropical inceptisols, with base satiration > 50% at pH 7.</td></tr>
<tr><td>2</td><td>Entic Hapludoll</td><td>Mollisol</td></tr>
<tr><td colspan="3">Clayey soils of little depth, locally colluvial. Rocks, debris, and rocky outcroppings present. pH < 7. This is a moister typical mollisol with entisol-like features.</td></tr>
<tr><td>4</td><td>Hapuldoll</td><td>Mollisol</td></tr>
<tr><td colspan="3">Soils are silty-clay to clayey, locally colluvial, and deep. Rocks are present. pH < 7. This is a typical moister mollisol.</td></tr>
<tr><td colspan="3">II. Sedimentary slopes</td></tr>
<tr><td>6</td><td>Hapludoll</td><td>Mollisol</td></tr>
<tr><td colspan="3">Soils are silty-clays, locally colluvial, and deep. pH < 7. THis is a typical moister mollisol.</td></tr>
<tr><td>9</td><td>Hapludoll and Udorthent</td><td>Mollisol and Entisol</td></tr>
<tr><td colspan="3">Colluvial soils that are silty and rocky, with little to very little depth. Numerous debris and rocky outcroppings. pH 6–7. Udorthents are moister orthents, which are loamy and clayey endisols with a regular decrease in content of organic matter with depth.</td></tr>
<tr><td>10</td><td>Entic Hapludoll</td><td>Mollisol</td></tr>
<tr><td colspan="3">Silty clay to clayey soils, locally colluvial, of little depth. Rocky. pH < 7. This is a moister mollisol with entisol-like features.</td></tr>
<tr><td>15</td><td>Vertic Hapludoll</td><td>Mollisol</td></tr>
<tr><td colspan="3">Deep clayey soils, locally colluvial. pH > 7. This is a drier mollisol with vertisol-like features.</td></tr>
<tr><td>17</td><td>Ustorthent and Entic Haplustoll</td><td>Entisol and Mollisol</td></tr>
<tr><td colspan="3">Colluvial silty-clay soils with very little depth. Rocks and rocky outcroppings are present. Ustorthents are drier orthents, which are loamy and clayey endisols with a regular decrease in content of organic matter with depth. Haplustolls are drier Mollisols, in this case with entisol-like features.</td></tr>
<tr><td>19</td><td>Paralithic Vertic Eutropept</td><td>Inceptisol</td></tr>
<tr><td colspan="3">These are colluvial clayey soils of little depth, with vertisol-like features. pH < 7. Eutropepts are tropepts, tropical inceptisols, with base saturation > 50% at pH 7.</td></tr>
<tr><td>20</td><td>Paralithic Vertic Ustropept</td><td>Inceptisol</td></tr>
<tr><td colspan="3">Clayey colluvial soils of little depth, with vertisol-like features. pH < 7. Ustropepts are similar to eutropepts, except that they form under drier moisture conditions.</td></tr>
<tr><td>21</td><td>Ustert and/or Vertic Ustropept</td><td>Certisol/or Inceptisol</td></tr>
<tr><td colspan="3">Very deep clayey vertisols and/or deep clayey inceptisols with vertisol-like features, locally colluvial. pH < 7.</td></tr>
<tr><td>24</td><td>Vertic Haplustoll</td><td>Mollisol</td></tr>
<tr><td colspan="3">Clayey soils with rocks, locally colluvial. Sandy and rocky outcroppings are present. pH > 7. These soils are drier mollisols with vertisol-like features.</td></tr>
</table>

(*continued*)

TABLE 8.1 (*continued*)

Code No.	Great Group Classification	Soil Order
III. Valley alluvium		
27	Ruventic Hapludoll and/or Tropofluvent	Mollisol and/or Entisol
Deep soils of variable texture, silty to silty clay. pH < 7; without $CaCO_3$. Moister river-deposited mollisol and entisols.		
28	Vertic Ustropept and Vertic Ustifluvent	Inceptisol and Entisol
Deep soils of variable texture, silty to silty clay. pH > 7; $CaCO_3$ present. Young river-deposited soils with little horizon development, and vertisol-like features.		
29	Typical Ustifluvent	Entisol
Deep soils with superimposed silty and sandy caps. Young river-deposited soils with little horizon development.		

had already been selected as "good" for maize by farmers, or at least good enough, given severe constraints on access to land for annual cropping.

From the viewpoint of growing maize, there is no problem with acidity in these soils. In fact, pH is a bit on the high side; high pH can reduce the availability of micronutrients such as zinc and copper. This is not a serious concern, however, given the pH values of the Jama Valley samples. Ample calcium and magnesium are available for maize growth. There are also high levels of potassium, which does not have any negative effect on maize yield or plant growth (maize is a "luxury" consumer of potassium, taking up high amounts into kernel development when available). High potassium levels likely come from the practice of burning fields before planting that I described earlier. Organic matter is high in most samples tested; this indicates that nitrogen is readily available for plant growth. Some fields are high in phosphorous, suggesting that manuring was practiced, or that the soil-forming material contained a relatively large amount of phosphorus-bearing minerals.

The overall favorable results of the soil analyses among different types of soils and areas of the valley highlight the role of effective moisture in maize productivity. In other words, effective moisture (considering both rainfall and evaporation, as well as soil drainage characteristics) rather than soil fertility appears to be the limiting factor for maize productivity. While farmers told me that maize prefers black soil, they emphasized the importance of rain for good yields. As the maize yield study will show (see below), in general fluvial lands and *brisa*-affected (seasonally foggy) nonfluvial lands are most productive for maize. These are lands with higher moisture (fluvial lands) and/or lower evaporation (*brisa* lands).

During the 1991 season of the yield study, Kennedy and I took short soil cores from fields we visited. We inserted the corer as deeply as possible into the field and matched the color of the soil extracted by the corer with values in a Munsell color chart (e.g., 7.5 YR 4/6; 2.5 Y 4/2). The color values provide information concerning the moisture and drainage characteristics of the soils. The number after the slash (chroma value) reflects the extent to which a soil is oxidized (well-aerated, well-drained) or reduced (poorly aerated, anerobic, less well-drained). Wetter, less well-drained soils are not necessarily bad for maize growth: in dry years, plants may do better in such soils than in better-drained soils. However, in wet years maize will do poorly in less well-drained soils. Maize is sensitive to the balance of water and air in soils, and prefers well-drained aerated soils (as long as rainfall is sufficient).

I have color data from 18 cores. These data were also interpreted for me by Randy Miles. Of the cores, 6 can be characterized as oxidized and well-aerated, while 12 are reduced and more poorly aerated. With a few exceptions, the better-drained soils were fluvial in origin, the less well-drained soils nonfluvial. Two of the better-drained soils were inceptisols of nonfluvial origin on steep terrain. I have ear length data for 4 of the fields with well-aerated soils and for 9 of the more poorly drained soils. Applying Student's *t*-test to these data, I found no significant difference in ear length. This is a small data set, however, and the data could not be separated by maize variety (which significantly influences cob length, as discussed below).

The color study results suggest that many farmers were utilizing lands that did not have optimal drainage/soil aeration for maize cultivation. As I discussed earlier, maize fields are tucked here and there around the Jama Valley today, squeezed in among pastures and plantations of coffee, cacao, plantain, and banana. Most maize is grown for household, rather than commercial, use. It is not surprising, then, that the crop is grown on less than ideally drained soils. More poorly draining soils are not necessarily a bad thing, however, especially in a dry year such as 1991. A number of the less than optimally drained fields produced large ears and relatively high yields.

Interpretations of soil test and soil color data are still preliminary. I do not have data from all parts of the valley, or all types of soils. Even given these limitations, however, the results are informative. Farmers selected lands for growing maize that were quite favorable for plant growth. Nutrients essential for growth were present, even abundant, in most fields tested. Most fluvial soils studied were well-drained, well-watered, and thus optimal for maize in both wet and dry years. Most nonfluvial soils selected for maize cultivation had poorer aeration. Since upland fields are largely solely dependent on rainfall for plant growth, more slowly draining soils are perhaps advantageous in dry years. Further study incorporating other variables affecting plant growth, such as field slope and aspect, and timing of rainfall, would clarify these patterns.

Analysis of Maize Data

As discussed earlier, three varieties of maize are common in the Jama River Valley today: *amarillo, cubano blanco,* and *hibrido.* During my fieldwork, I noticed that *amarillo* ears appeared to be smaller than the other varieties. Farmers also told me that there were differences in how these varieties of maize produced under different conditions of moisture and soil. I decided, therefore, to examine the agricultural data by maize variety. The disadvantage of this approach is that the number of cases is reduced for some of the analyses, as will be seen below.

When I reviewed my field data and classified each data set by maize variety seven groups were produced: *amarillo* (A), mixed plantings of *amarillo* and *cubano blanco* (AC), mixed plantings of *amarillo* and *hibrido* (AH), mixed plantings of all three varieties (ACH), *cubano blanco* (C), mixed plantings of *cubano blanco* and *hibrido* (CH), and *hibrido* (H). Table 8.2 shows the average ear length, in cm, for

TABLE 8.2 AVERAGE EAR LENGTH BY VARIETY, ALL DATA SETS

Variety	Length	Number(*N*) of Fields Studied
A	16.3	12
AC	16.7	7
AH	15.4	8
ACH	17.5	8
C	17.9	36
CH	17.5	20
H	17.1	31

each group. As the table illustrates, *amarillo* is grown less frequently today than *cubano blanco* and the hybrids.

When *amarillo* is planted with other varieties, the smaller *amarillo* ears reduce average ear length for the sample. A Student's *t*-test comparing ear length from plantings of pure *amarillo* and all mixed plantings containing this variety (AC, ACH, AH) resulted in a 0.76 probability that differences in length were due to chance (2-tailed test, variance not the same). In other words, when it was impossible to consider A, AC, ACH, and AH separately because of small sample sizes, I could combine them into a mixed *amarillo* group (AM).

As discussed in Chapter 5, for the first years of the project, I estimated the *potential yield* of maize in a field by taking a series of measurements. I measured the distances between rows and *matas* (planting groups) for a section of the field, counted the number of ears for a sample of *matas,* and then used these data to calculate the density of ears in the field. That figure, in turn, could be converted to ears/ha, allowing me to compare the productivity of different fields. I now think that this method overestimates actual yields. I came to this conclusion because potential yields are always much larger than yields as *recalled* by farmers for the 11 cases for which I have both for the same fields. One might be able to compare the two if there was a consistent relationship between them, but this is not the case.

I believe it is still useful, however, to examine how potential maize yields vary between alluvial and nonalluvial soils in the valley, by elevation, and among Stratum I, II, and III. The only ears we counted in fields were those that were clearly developing normally—already elongated with obvious cob development. In local terms, most were between *choclo* (green corn) and *mazorca* (dried corn) stage when I studied them. When a farmer picked 20 ears for me to measure, more often than not most were fully developed. These observations suggest that final yields were reduced by factors occurring late in the growing season, after my visits to fields. Too early cessation of rain, damage to drying ears from insects, and toppling of plants by wind were all mentioned to me by farmers. Thus the potential yield data can perhaps be considered the "best-case" scenario for maize productivity—a way to model optimal years. I'll return to this point later.

For the final field season of the project, I asked crop owners, or their family members, for the yields of particular fields (*recall yield* data). At the same time, I asked for the size of the field, its approximate location, whether it was flat or sloping, and how the maize was spaced. Although the owners could not always remember how the field was planted (or sometimes the family member I interviewed did not know), they answered all my other questions without hesitation. All members of the household knew about field size and harvest, facts often also known by neighbors and relatives.

These methodological issues considered, let us examine the hypotheses with the maize data.

H1. That alluvial lands are more productive than nonalluvial lands, and productivity differs between major orders of nonalluvial soils

The results of this comparison using ear length data are presented in Table 8.3. A statistically significant difference exists in average ear length on alluvial and nonalluvial soils for *cubano blanco,* and the difference is nearly significant for hybrid maize. In both cases, ears are significantly larger on alluvial soils. No significant differences were found for *amarillo* or for mixed plantings of *amarillo* and other maize. Note, however, that when all maize data are considered together (increasing sample sizes), maize ears are significantly larger when grown on alluvial soils.

For the above analysis I grouped all alluvial and all nonalluvial data, regardless of soil type. I found no significant difference between maize yields on the moister, mollisol-derived alluvium of Stratum III and the drier, less developed alluvium of Stratrum I.

Table 8.4 presents *potential yield* data, in ears/ha, of alluvial and nonalluvial soils for all maize varieties. All *amarillo* data are combined, since few data sets exist for this variety. Fields of *amarillo* produce significantly more ears/ha on alluvial than nonalluvial soil. This pattern also holds for *hibrido,* at nearly significant levels. *Cubano blanco* and mixed plantings of *cubano* and *hibrido* show no significant differences in productivity, however. When all data sets of potential yield are combined, the hypothesis is supported.

What is the impact of crop spacing on potential yields in alluvial and nonalluvial settings? If, for example, farmers planted maize further apart on alluvial soils, so that beans, chile peppers, and other crops requiring high moisture could be interplanted with the maize, this could draw the yields of maize down. Planting maize further apart in upland soils, to increase the access of plants to water, could reduce yields per hectare. To explore these possibilities, I looked at whether there was a significant difference in maize spacing (measured as distance between rows × distance between matas) between alluvial and nonalluvial plots. The spacing product for alluvial fields averaged 1.25 ($N = 31$); for nonalluvial, 1.59 ($N = 36$) for all varieties of maize combined. This is close to a significant difference (t-test, 2-tailed, variance not the same; 0.10), with maize spaced further apart on average in nonalluvial plots. Thus, part of the difference between potential yields on alluvial and nonalluvial plots results from less dense plantings on nonalluvial soils.

Several farmers told me that the further apart maize is planted, the larger the ears will be. To explore this possibility, I graphed cob length versus the spacing product for alluvial and nonalluvial data sets. I then fitted a Method Two regression line to each graph. A good fit (equal numbers of points above and below the line) indicates that the direction and slope of the line can be interpreted. There was no clear relationship between these variables for alluvial plots, but for nonalluvial plots (Figure 8.5), there was a slight positive relationship. The benefits of wider spacing of maize *matas* on ear development are modest. Optimal planting distances are similar on alluvial and nonalluvial soils: between 1 × 1 m and 1.25 × 1.25 m spacing.

TABLE 8.3 COMPARISON OF EAR LENGTH ON ALLUVIAL AND UPLAND (NONALLUVIAL) SOILS*

	Average Length Alluvial	*N*	Average Length Nonalluvial	*N*	*t*-test
A	16.7	4	16.1	8	0.38
AC	—	0	16.7	7	—
ACH	17.1	4	17.9	4	0.28
AH	15.8	2	15.2	6	0.35
C	18.8	10	17.6	26	0.01
CH	17.7	6	17.5	14	0.40
H	17.6	11	16.9	20	0.13
AM group	16.7	10	16.3	25	0.35
All maize	17.7	37	17.0	86	0.03

*Sample averages and *t*-test results (probability of differences being due to chance) are presented for each maize group; *t*-test is 1-tailed, variance unequal

TABLE 8.4 POTENTIAL YIELD IN EARS/HECTARE*

	Average Yield Alluvial	*N*	Average Yield Nonalluvial	*N*	*t*-test
All AM	28,467	6	15,780	5	0.04
C	19,450	8	21,689	18	0.25
CH	28,933	6	34,067	3	0.32
H	29,278	9	24,163	8	0.13
All maize	26,657	30	22,434	35	0.06

**t*-test is 1-tailed, variance unequal

Table 8.5 illustrates differences in yield between alluvial and nonalluvial plots as *recalled* by farmers. There is a significant difference for all maize varieties, combined, *against* the hypothesis, and a nearly significant difference for the *amarillo* group and for *cubano blanco,* also against the hypothesis.

As I just discussed, there is a slightly positive relationship on nonalluvial plots between wider spacing of maize and higher potential yields, and also between wider spacing and longer ears. This relationship also holds for recall yield data, for both alluvial and nonalluvial fields (Figure 8.6). If increased water availability leads to better kernel development, fewer ears would be discarded during harvest, increasing the recall tally accordingly.

As discussed earlier, cultivation practices being equal, nonalluvial fields located on relatively deeper, more fertile mollisols should produce better than fields on nonalluvial inceptisols, especially thin soils on steeply sloped terrain and those that shrink and swell seasonally. Results of this comparison, the second part of H1, are presented in Table 8.6.

The hypothesis is strongly supported by the ear length data (except for *hibrido,* which produces equally well on both types of soils), less strongly by the potential yield data, and rejected by the recall data. Overall, the maize yield results seem to sup-

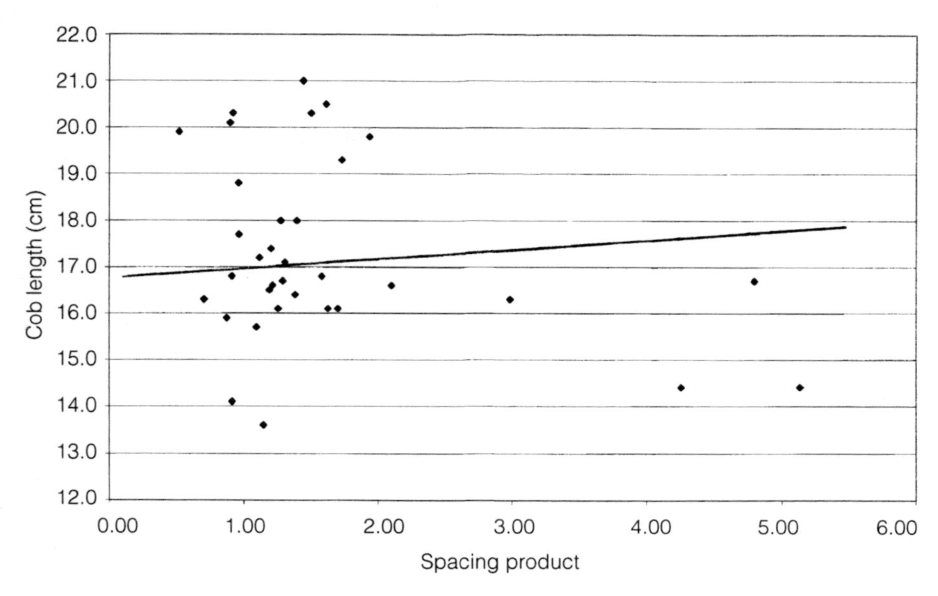

Figure 8.5 Graph of cob length versus plant spacing product (distance between rows × distance between matas*)*

TABLE 8.5 AVERAGE RECALL YIELDS ON ALLUVIAL AND NONALLUVIAL PLOTS*

	Alluvial Average	***N***	**Nonalluvial Average**	***N***	***t*-test**
AM	4,925	4	8,400	15	0.07
C	5,088	6	7,310	17	0.07
All maize	5,058	13	8,597	49	0.001

**t*-test is 1-tailed, with unequal variance

port the hypothesis that mollisols are more productive than inceptisols in the Jama Valley. Maize yields on mollisols are in fact close to those obtained on alluvial soils.

Finally, I compared the moister udolls and drier ustolls *within* the mollisol group, to see if maize yielded better on the moister mollisols, which occur in Stratum III and in part of Stratum II, than on the drier ones, which occur near the coast. Recall data supported the hypothesis; potential yield showed no significant difference; and cob length data supported the opposite relationship. Given the inconclusive nature of these results, I will consider moister and drier mollisols equal for maize productivity. Some maize varieties yield very well on the drier coastal soils.

To summarize, there is considerable support from the maize study for Hypothesis 1, that alluvial lands are more productive for agriculture than nonalluvial lands, and that productivity also differs between major orders of nonalluvial soils. *Cubano blanco* and *hibrido* produce significantly larger ears on alluvial than on nonalluvial soils, for instance. While *amarillo* did not show this difference, the relationship holds for all the maize ear length data combined. *Amarillo* produces

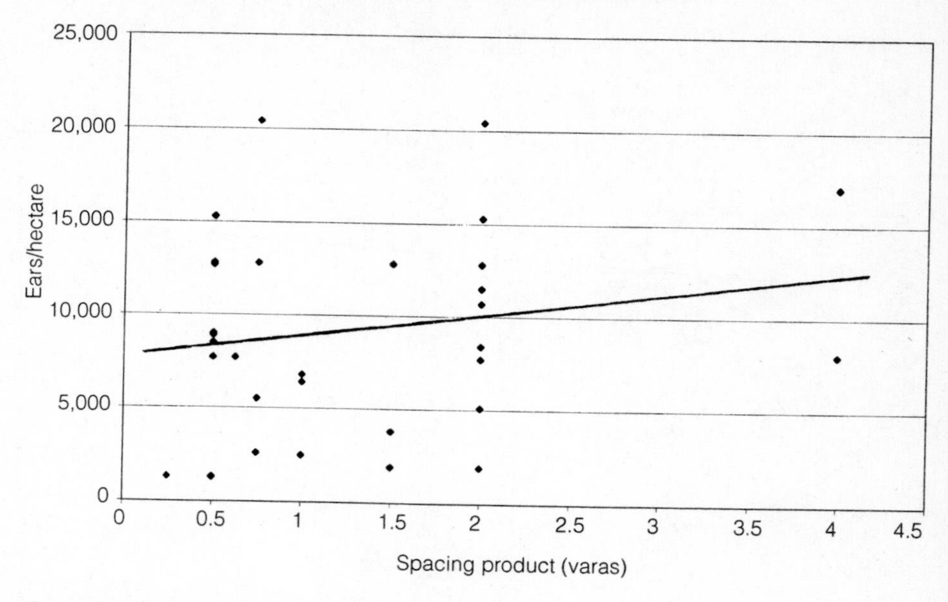

Figure 8.6 Maize recall yields versus mata spacing

TABLE 8.6 COMPARISON OF NONALLUVIAL MAIZE YIELDS BY SOIL ORDER*

	Mollisol Average	*N*	Inceptisol Average	*N*	*t*-test
Cob length					
All maize	17.4	50	16.5	35	0.004
AM	16.9	12	15.8	13	0.05
C	17.9	18	16.8	8	0.03
CH	17.7	9	17.0	5	0.12
H	16.9	11	16.9	9	0.49
Potential yield					
	24,096	23	19,250	12	0.11
Recall yield					
	6,757	24	10,364	25	0.01

*t-test is 1-tailed, with unequal variance

significantly more ears/ha on alluvial than nonalluvial soil, as does *hibrido,* at nearly significant levels. Maize yields as recalled by farmers do not support the hypothesis. Nonalluvial soils are not all equal for maize production. Soils in the mollisol order (eight types) produce longer maize ears than soils in the inceptisol order (four types), with the exception of *hibrido,* which produces equally large ears on either type of soil.

H2. That agricultural productivity in the uplands (nonalluvial soils) increases as elevation increases

I first explored this relationship (or lack of relationship) by graphing cob length versus elevation, for all varieties of maize and all parts of the valley. While it was possible to fit a regression line to this graph, no strong relationship between the two variables was shown.

Since my personal observations of the *brisa* effect were made in the area southwest of San Isidro, I decided to graph cob length data by stratum. No relationship existed between the variables for Stratum I and Stratum II, but Stratum III showed a positive relationship, based on a fairly good fit (Figure 8.7). I obtained the same result when I graphed only the data from fields located southwest of San Isidro.

I then graphed *potential yield* versus elevation for all nonalluvial maize data, and found no clear relationship between these variables, the same result as the comparison of elevation and ear length. However, when I graphed the data by stratum, there was again a clear trend for increased yield in ears/ha with elevation for Stratum III, as well as for the area southwest of San Isidro where the *brisa* effect is the most pronounced.

Finally, I graphed *recall yield* data for all fields, regardless of type of maize, against elevation for the whole valley. While there is considerable variability in yield by elevation, I found a fairly good fit in the Method Two regression line between increased elevation and higher recall yields on nonalluvial soils. Among the individual varieties of maize, the fit is strongest for *amarillo* and *hibrido*. Plant spacing does not seem to influence the results of examining recall or potential yield by elevation.

To summarize, Hypothesis 2, that productivity on nonalluvial soils increases at higher elevations, is only weakly supported for the valley as a whole by the maize data. Only the *recall yield* data show a valleywide trend for increased yields on higher-elevation nonalluvial soils. Maize produces larger ears at higher elevations only in Stratum III. This is especially true for *cubano blanco* and *hibrido*, with a weak positive relationship for *amarillo*. Similarly, potential yields for all varieties combined increases with elevation only in Stratum III. Thus the second hypothesis is supported strongly only for Stratum III.

H3. That there are differences in agricultural productivity among Stratum I, Stratum II, and Stratum III

Because of the low numbers of cases per stratum for *amarillo* plantings, I examined the relationship of ear length and stratum using combined *amarillo* data (AM). The other varieties and groups can be examined separately. Sample means are shown in Table 8.7. For the *amarillo* group, sample means are very similar among the three strata, indicating little variability by stratum for this traditional maize group. There is much more variation among the means for *cubano blanco*, with the largest ears produced in Stratum I. *Hibrido* maize cob development is quite similar in Stratum I and II, but cobs are smaller on average in Stratum III. Finally, mixed plantings of *cubano blanco* and *hibrido* parallel the trend for *cubano*.

Table 8.8 summarizes *potential* maize productivity by stratum. There are very few cases for Stratum II ($N = 8$), making it impossible to examine the data for this stratum by maize variety. Stratum I and Stratum III are very similar in overall potential maize production, and exhibit higher yields per hectare than Stratum II. All in all, potential maize productivity as measured by ears/ha is quite similar around the valley. If the few data from Stratum II are representative, this is the least productive

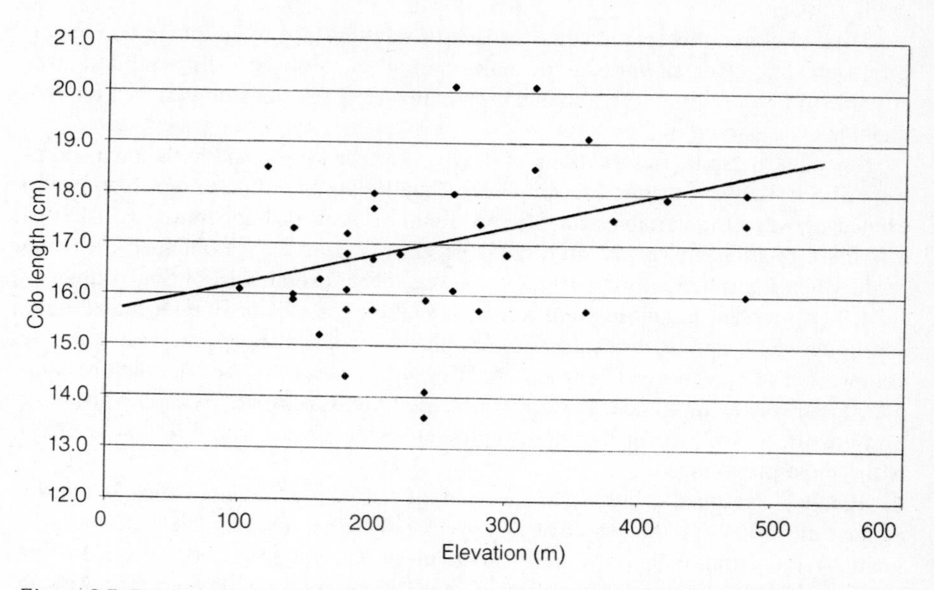

Figure 8.7 Relationship between elevation of a field and cob length for Stratum III cases (all maize varieties)

TABLE 8.7 COMPARISON OF AVERAGE COB LENGTHS, BY VARIETY, AMONG STRATUM I, II, AND III

	Stratum I Average	*N*	Stratum II Average	*N*	Stratum III Average	*N*
AM	16.7	6	16.5	17	16.3	12
C	18.8	12	16.9	6	17.7	18
CH	18.0	5	17.2	3	17.5	12
H	17.6	7	17.5	4	16.9	20

TABLE 8.8 AVERAGE POTENTIAL YIELDS, IN EARS/HECTARE*

	Stratum I Average	*N*	Stratum II Average	*N*	Stratum III Average	*N*
AM	—	3	—	2	23,800	6
C	19,440	10	—	3	23,646	13
CH	—	3	—	1	29,060	5
H	—	2	—	2	24,892	13
All maize	25,105	18	19,663	8	24,841	37

*Averages not calculated for three or fewer cases

zone, which is not surprising given the lack of alluvial soils there, the rugged topography, and the comparatively smaller area.

I also compared average yields of maize as recalled by farmers among the three strata of the valley, with the results shown in Table 8.9. Yields are lower in Stratum I than in either Stratum II or Stratum III, which are fairly similar.

In summary, Hypothesis 3, that agricultural productivity differs among Stratum I, II, and III, is difficult to evaluate. Looking at maize ear length, *amarillo* produces ears of similar length everywhere in the valley, while *cubano blanco* produces its longest ears in Stratum I. Potential yield as measured by ears/ha in the field is quite similar around the valley, with Stratum II being the least productive zone. Finally, yields as recalled by farmers are lower in Stratum I than in either Stratum II or Stratum III, which are similar in average yield. All in all, while there is variation in the various data sets among the valley strata, there is no consistent patterning. The small sample sizes for Stratum II—arguably the most difficult area to farm—may also contribute to the problem.

This inconsistency may also be explained, in part, by the maize varieties grown. *Amarillo* has a long history in the region and is well adapted to conditions throughout the valley. *Cubano blanco* is Caribbean in origin and is thus well adapted to dry growing conditions. The modern hybrids also contribute to the success of maize agriculture today, producing equally well in inceptisol and mollisol soils, and in dry (Stratum I) conditions.

I have also come to believe that Hypothesis 3 was not especially well framed to begin with. In other words, there may be better ways to examine valleywide variation (or the lack of it) in maize productivity than by using the survey strata. Comparing maize yields on alluvial soils of the lower valley to those of the middle valley, for example, revealed that no significant differences existed in any of the measures. This translates into no differences within the most productive maize cultivation zone, the annually renewed river alluvium, despite some variation in pH and soil texture among alluvial soils (the coastal alluvium being more alkaline than that of the middle valley, and with vertisol-like characteristics).

Nonalluvial soils are not all equal in maize yields, however, and this has important implications for modeling maize productivity within the valley. The hypothesis that mollisols outproduce inceptisols is supported by the cob length data and potential yield data, and is rejected by the recall data (which supports the opposite relationship). Overall, I would accept the hypothesis, since maize yields on mollisols are in fact close to those obtained on alluvial soils. The Jama Narrows (Stratum II) is dominated by inceptisols, and these soils are on steep terrain. Further, alluvial deposits are lacking. All these factors suggest that the Narrows is the least attractive part of the valley for maize cultivation. The lower (Stratum I) and middle (Stratum III) valley zones are roughly equal in area extent of mollisols. Much of the northeastern part of the middle valley is dominated by inceptisols, however, with large expanses of mollisols located to the southwest of San Isidro. The latter region is notable for higher maize productivity with increased elevation and a pronounced *brisa* (fog-moisture) effect; cultivation of CCP is in fact possible on ridgetops in this zone, as I discuss further below. The San Isidro/southwest region of the middle valley, not Stratum III as a whole, may thus have an advantage in maize productivity: a concentration of better nonalluvial soils, and large pockets of river alluvium.

TABLE 8.9 AVERAGE RECALL YIELDS, IN EARS/HECTARE[1]

	Stratum I Average	N	Stratum II Average	N	Stratum III Average	N
AM	—	2	8,462	13	8,100	4
C	—	1	7,600	4	5,810	10
CH	7,800	4	—	1	—	3
H	5,617	6	—	2	12,342	7
All maize	5,344	13	8,854	24[2]	8,216	25[2]

[1]Averages not calculated for cases with three or fewer data points.
[2]Includes data sets for which variety is unidentified.

FINAL THOUGHTS ON THE MAIZE STUDY I have presented three types of maize data—ear length, potential yield estimates (ears/ha), and yield as recalled by farmers (*fanegas/cuadra*, converted into ears/ha)—and used them to evaluate the major hypotheses of the study, with the results summarized above. After finishing these analyses, I wondered how the results might have been influenced by the fact that I collected these data over five different growing seasons. Table 8.10 illustrates how the ear length data, for example, break down by year. Except for six fields, all the 1991 data are nonalluvial; there is a more even mixture of alluvial and nonalluvial in the other years. There is a fairly even mix of maize types for each year.

The 1983 growing season produced the longest ears of the study period. This is not surprising: this was the year of the strongest El Niño event since 1925. The rainy season started early, in late 1982, and continued for over a year. When fieldwork resumed in 1988, maize ears were smaller, and continued to decline in size during 1989 and 1991. A number of farmers spoke of drought conditions during the 1991 growing season. The fact that most of the 1991 data are from upland (nonalluvial) fields probably accentuates this trend toward small ears. There is a significant difference between cob length data from the wettest year (1983) and the (presumably) driest (1991) year (*t*-test, 2-tailed; 0.02). There is also a significant difference between the two largest sets of data, 1988 and 1991 (*t*-test, 2-tailed; 0.002).

I compared alluvial data sets to nonalluvial for 1988 and 1989, the two largest sets of data with good representation of these soil types. I found that maize grew larger on alluvial soils in 1988 (*t*-test; 0.07), but not in 1989 (*t*-test; 0.31). When I graphed cob length versus elevation for 1991, the year with the most nonalluvial data, there was no clear relationship between increased elevation and increased cob length for the data as a whole. A positive relationship did exist for the combined *amarillo* group, however.

In summary, there are too few data sets to look at all the hypotheses separately by year and by type of maize. Looking at some of the variables this way shows that the trends discussed earlier for cob length hold up, for the most part. This suggests that the data from various years can be combined. The study spanned both very wet and dry years, capturing the natural variability of rainfall in this transitional lowland tropical forest zone.

Finally, a few comments are in order about comparing and using recall and potential yield data. Remember that yields as recalled by farmers are lower than the estimated potential yields from field measurements for cases for which I have both types of data. This suggests that my method for determining potential yield did not

TABLE 8.10 AVERAGE EAR LENGTH DATA, BY YEAR (ALL TYPES OF MAIZE COMBINED)

Year	*N*	Average Ear Length
1982	4	16.8 cm
1983	10	18.3 cm
1988	39	17.8 cm
1989	23	17.2 cm
1991	46	16.6 cm

take into account late-season factors, such as poor kernel fill, that affected harvests. Farmers told me that 6–7 fenegas/cuadra was a very good harvest, 4–5 fenegas a good harvest, 3 fenegas or so a "regular" harvest, and less than that a poor harvest. If I classify the 65 potential yield data sets this way, 74% fall far *above* the level for very good, with 15% very good, 2% good, 2% regular, and 8% poor. Again, this suggests that the potential yield results are inflated.

Should the *potential* yield data be dropped from the study as unrealistic, or used to model optimal harvests for the valley? I am tempted to keep the data in because they are patterned in ways that seem logical: fields containing *amarillo* produce significantly more ears/ha on alluvial than nonalluvial soil, a pattern also true for *hibrido,* at nearly significant levels, and for the combined maize data; yields increase with elevation in Stratum III; and maize productivity as measured by ears/ha in the field is quite similar around the valley, with Stratum II being the least productive zone.

Maize *recall* yield data, on the other hand, may represent the low end of maize productivity in the valley. Remember that most of these data were collected during drier years, and represent mostly nonalluvial plots. If I compare the maize recall data (62 cases) to farmers' expectations of yield quality, 48% of the harvests for which I have data would be considered poor, 23% regular, 15% good, and only 10% very good. An additional 5% produced harvests greater than 7 fenegas/cuadra, while many of the poor harvests netted considerably fewer than 3 fenegas of maize per cuadra. Again, these results support the interpretation that dry conditions late in the growing season adversely affect harvests, especially on nonalluvial soils.

Analysis of Yuca Data

To begin the analysis of yuca yield in the Jama Valley, I reviewed information from field interviews and classified the cases by yuca variety. As discussed in the beginning of this chapter, four types of yuca are grown in the Jama Valley today: *yuca de tres meses* (3-month yuca), *espada blanca* (6-month yuca), *yema de huevo,* and *yuca negra* (both 1-year yuca). Table 8.11 summarizes the yuca weight data I collected by length of maturation of the variety. As the table illustrates, average root weights are similar for all varieties, and for mixed plots of yuca. A *t*-test comparison of the varieties with the lowest and highest averages, 1-year and 3-month yuca, gave a result of 0.68 (2-tailed test). These results indicate that yuca data can be examined as a single unit, rather than by variety. This significantly increases sample sizes.

As discussed in Chapter 5, I collected two types of data to represent yuca yield in the Jama Valley: root weight per plant (lbs/*mata*) and potential yield estimates (*matas*/ha × lbs/*mata,* to give lbs/ha). I will use these data to address the hypotheses.

TABLE 8.11 YUCA ROOT WEIGHT DATA, IN POUNDS OF UNPEELED ROOTS

Variety	**Average Weight**	**Number of Plots**
Three-month	9.11	18
Six-month	10.18	8
Year	7.83	3
Mixed plots	9.37	8

Unfortunately, there are too few data from Stratum I and II (five sets all together) to draw comparisons among the valley strata, so Hypothesis 3 cannot be evaluated.

There are two primary sources of error in the data I have for modeling yuca productivity in the valley. First, I must assume that the weight of roots from three or four plants represents the average potential root development in a field of yuca. Farmers always selected "good" plants to pull up (i.e., ones that they judged were ready for use), but in many cases I was told that the yuca would be left in the field to develop further. This suggests the root weight data may be on the low side. Second, I must assume that all the plants living in a field when I studied it would go on to full maturity. This is less problematic than such an assumption about maize, since there seems to be no insect or animal predation of yuca, it is not bothered by wind, and its response to failing rain is simply to go dormant. It thus seems reasonable to assume that most *matas* of yuca will be harvested as needed before the roots rot in the ground.

I have two sets of *recall* yield data for yuca fields harvested in their entirety for sale. Both are nonalluvial plots in Stratum III. Yields were 35,400 and 17,400 lbs/ha. These figures are well within the range of the potential yield calculations; 35,400 lbs/ha is in fact close to the average potential yield of yuca.

H1. That alluvial lands are more productive for agriculture than nonalluvial lands, and that productivity also differs between major orders of nonalluvial soils

Results of comparing yuca yield on alluvial and nonalluvial soils are presented in Table 8.12. Weight of yuca roots per mata is not significantly higher on alluvial soils. When estimated numbers of plants per hectare are factored in, total potential yield is also not significantly higher on alluvial soils. Yuca clearly yields equally well on both classes of soils. There is also no difference between production on alluvial lands of the upper and lower valleys; both alluvial areas are equally productive for yuca.

I also compared yuca production on the two orders of soils represented in the nonalluvial plots in the valley (Table 8.13). Yuca root production is not significantly heavier on mollisols than on inceptisols. In fact, when planting density is factored in, plots on inceptisols outproduce plots on mollisols, at a level of near significance. Yuca productivity is clearly not as influenced by soil type as is maize.

H2. That agricultural productivity in the uplands (nonalluvial soils) increases as elevation increases

I graphed elevation versus yuca root weight for all plots on nonalluvial soil and found no apparent relationship between elevation and per plant yield. Similarly,

TABLE 8.12 COMPARISON OF YUCA YIELDS ON ALLUVIAL AND NONALLUVIAL SOILS*

	Alluvial Average	***N***	**Nonalluvial Average**	***N***	***t*-test**
Pounds/mata	8.68	14	10.20	26	0.22
Pounds/ha	36,028	14	38,620	25	0.37

**t*-test is is 1-tailed with unequal variance

TABLE 8.13 COMPARISON OF NONALLUVIAL YUCA YIELDS BY SOIL ORDER*

	Mollisol Average	***N***	**Inceptisol Average**	***N***	***t*-test**
Pounds/mata	10.51	17	9.60	25	0.36
Pounds/ha	30,341	16	53,339	9	0.09

**t*-test is 1-tailed, with unequal variance

when I graphed elevation versus potential yield in lbs/ha, there was no apparent relationship between these variables.

To explore the impact of spacing differences on yield calculations, I first calculated the average spacing for yuca (expressed as the product of the distance between rows and the distance between *matas*) grown on alluvial and nonalluvial plots (Table 8.14). As the table shows, there is no significant difference between how yuca plantings are spaced on these soils.

I then graphed spacing versus root weight for all plots, and found a positive relationship between these variables. In other words, with considerable variation, the wider spaced the plants, the larger the root weight. Farmers did not mention this relationship during interviews; by contrast, I was told repeatedly that this was the case with maize. Since yuca and maize are commonly interplanted in alternating rows, spacing in many yuca plantings is likely a by-product of spacing decisions made with maize in mind.

Finally, I graphed yuca spacing versus elevation, and found no apparent relationship. In other words, decisions about how densely to plant yuca did not relate to elevation.

Conclusions for the Yuca Study

There is great variability in the yuca yield data I collected during the Jama project. For example, average root weight, measured from three or four plants uprooted by the crop owner, ranged from 1.75 lbs/mata to 24.8 lbs/mata, for an overall average of 8.68 lbs/mata for alluvial plots and 10.2 lbs/mata on nonalluvial soils. Given the range of variation, this is not a statistically significant difference, nor is the difference between potential yield per hectare on alluvial and nonalluvial soils statistically significant. Further, there is no apparent relationship between elevation and yield; yuca does not seem sensitive to the *brisa* effect. It also produces well on both mollisols and inceptisols. Clearly yuca produces equally well today under many different soil and moisture conditions in the Jama River Valley.

TABLE 8.14 COMPARISON OF SPACING ON ALLUVIAL AND NONALLUVIAL YUCA PLOTS*

	Alluvial Average	*N*	Nonalluvial Average	*N*	*t*-test
Spacing product	3.28	14	3.21	25	0.95

**t*-test is 2-tailed, with unequal variance

Interviews with farmers give some additional insight into these results. As discussed earlier in the chapter, how long yuca takes to mature—to produce good, heavy roots—depends on the rain. Only 10–12 lbs/mata is expected if yuca is harvested during its first growing season. Yuca left in fields into a second rainy season, or for up to 18–24 months, depending on the variety, may yield 25–30 lbs of roots/mata according to local farmers.

I graphed root weight against yuca age at harvest for yuca plots planted in December or January (Figure 8.8). There is no clear relationship between age and root size, although there is some tendency for heavier roots to come from older plants. Interestingly, for the few data points I have of old plantings (20–23 months), anticipated high yields did not materialize. Average yield for these oldest plants fell far below 25–30 lbs/plant. All the yuca I measured had been left in fields beyond the end of its first rainy season (i.e., beyond 5 or 6 months), and there is a slight tendency for root weight to increase with the onset of the next rains (i.e., for plants aged 12–13 months). Overall the weight data I collected in the field (8.68 lbs/mata on alluvial, 10.2 lbs/mata on nonalluvial soils) fell in the range of weights anticipated by farmers for yuca harvested during its first growing season.

These results suggest, then, that yuca productivity in the Jama Valley for the years of the field study, like maize productivity, did not come up to farmers' expectations. This is especially evident in the failure of root weight to double for *matas* left for two full rainy seasons. Harvest early in the second rainy season did result in somewhat higher weights. Overall, measured root production was not far under expected production for yuca grown for one rainy season. The underlying causes of the large variation in root weight of individual plantings remain obscure. Sampling error from the small numbers of roots measured per field is a likely contributing factor.

LAND-USE STUDY RESULTS

As I discussed in Chapter 5, in addition to collecting maize and yuca yield data for plots in the valley, I also sketched on aerial photo overlays the location of fields used for annual cropping, plantation crops (CCP), and pasture (see Figure 5.7). I was especially interested in the location of maize fields relative to CCP. I reasoned that since coffee, cacao, plantain, and banana are demanding of water and well-drained, fertile soils, the location of such plantations would indicate lands especially productive for maize (but not used today for maize, since it is not an important cash crop). Plantation agriculture is indicated on the topographic maps for the study area, but I also wished to "ground truth" this information.

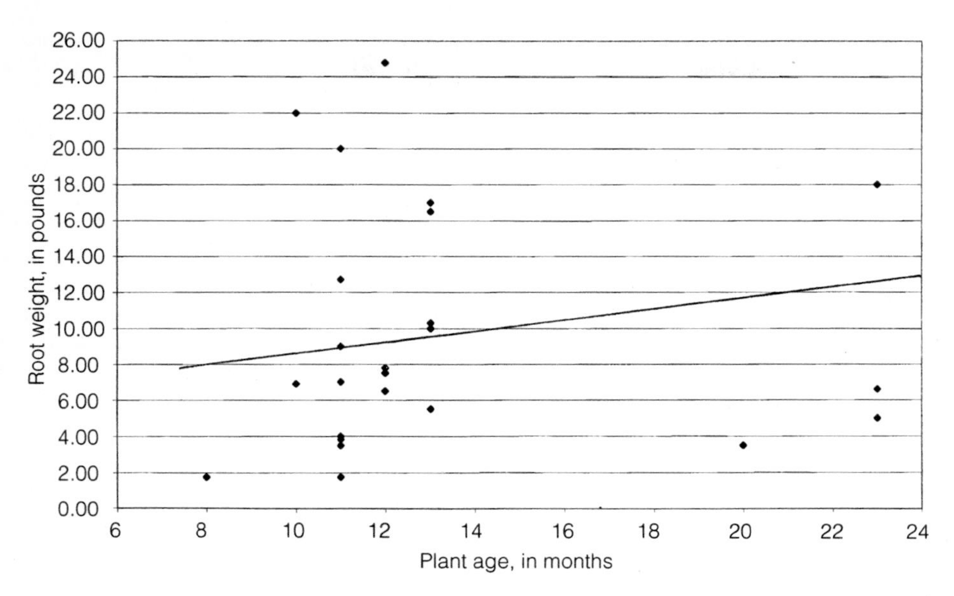

Figure 8.8 Relationship between the age of yuca plants and the weight of harvested roots

Land-use survey in the coastal zone (Stratum I) and the southwestern sector of the Narrows (Stratum II) confirm the land-use patterns shown on topographic maps. Nonfluvial lands are in secondary forest or pasture, not in plantation crops. The only CCP plantings in this sector of the valley are on the Jama River alluvium, from the town of Jama upstream to the beginning of the Narrows, and such plantings are not extensive. Maize is also planted on alluvial soils along the Jama River, as well as along or just above small streams in the lower valley, such as the Boca de Bigua region along the Río Mariano. I also observed maize plantings in nonalluvial settings in the lower valley, such as in the Campamento area west/southwest of the town of Jama. Modern land use thus parallels soil data for the lower valley: There are extensive tracks of upland mollisols in the lower valley that can be used for maize and yuca cultivation, but they are not moist enough for CCP. While these soils are drier than mollisols of the middle valley, recall that the maize yield study suggests they yield equally well or better (for some maize varieties) than moister mollisols.

I also made observations in Stratum II to the north of the Río Jama, both close to the river and further northeast, along the Estero Sálima. There is some CCP along the lower reaches of the E. Sálima, as it approaches the Río Jama, but none in the upper reaches or in the upland areas of Stratum II. This is in agreement with land-use classification on the topographic map. The uplands of the Narrows are likely too steep, and the soils too thin, for plantation agriculture. Maize fields are scattered in the lower slopes above secondary streams, and in higher areas where there is road access. Soils are predominantly inceptisols; there are no alluvial deposits along the Jama River in the Narrows.

Working out of San Isidro, I made a series of land-use observations in Stratum III. Early in the project I focused on the areas to the west, south, and east of San Isidro, and confirmed that CCP plantations occur commonly in the uplands on ridgetops between stream drainages, as well as in alluvial pockets along the Jama and its major tributaries. Upland maize fields were typically located below the ridgetop CCP areas, with pasture (relatively little of it) located below that and above riverside CCP plantings (see Figure 5.7). I suspect that pasture areas were formerly planted in maize. This is the area where I first learned of the *brisa* effect. The presence of healthy CCP plantings, as well as the limited maize data I have analyzed, confirms that these moist ridgetops are productive croplands.

Moving east toward Eloy Alfaro, as well as north/northeast of San Isidro (north of the Jama River), there is no ridgetop CCP, only secondary forest and pasture, with some CCP along the river. Inceptisols replace mollisols in these areas, and there is relatively little maize cultivation today. Upland plantation areas depicted on the topographic map north of Eloy Alfaro do not exist today; this area is dry and deforested. There is some ridgetop CCP just outside our study region, in one of the pockets of mollisols in this largely inceptisol area.

Finally, moving downriver from the Boca de Calada area (confluence of the stream along which San Isidro is located and the Río Jama) toward Pechichal and the beginning of the Narrows, there is limited CCP and maize cultivation along the river, and large expanses of pasture on nonalluvial soils, with no CCP. This is an inceptisol area.

My study of land-use patterns in the valley helped me become familiar with the land and the distribution of resources across it. There are many reasons why a given parcel of land may be used for a particular purpose, or why land may lie seemingly unused. Social and economic factors are of paramount importance today: Many people either have no access to land or only to marginal areas as small holdings are consolidated into larger parcels, and cattle pastures replace crop production. Based on my observations, conversations with knowledgeable local people, and study of soil and topographic maps, it seems clear, however, that underlying these considerations are some important environmental factors that are useful for considering the prehistoric situation. The locations of CCP plantings in the uplands, in combination with the distribution of alluvial soils, help delineate areas productive for maize in the past.

THE RECONSTRUCTED AGRICULTURAL LANDSCAPE

A Working Model and Predictions

Based on results of my study of maize and yuca agriculture in the valley, I propose a working model for agricultural productivity that defines three broad classes of maize lands, from highest to lowest yielding: river alluvium (all equal), mollisols (two levels of productivity), and inceptisols (all equal). Higher-elevation mollisols in Stratum III, especially the region southwest of San Isidro, are modeled as more productive for maize than other mollisols. Figure 8.9 illustrates the distribution of these classes of lands in the valley.

For the final project monograph, we will produce GIS overlays that combine maize and yuca yields with soil, elevation, slope, and aspect data to model productivity more precisely, and permit analysis of archaeological site locations in terms of landscape variables. For now, a few observations from the working landscape model may help us understand prehistoric plant-people interrelationships in the valley.

First, agricultural productivity in the valley is not uniform today, and would not have been uniform in the past. Most of the Jama Narrows (Stratum II) is character-

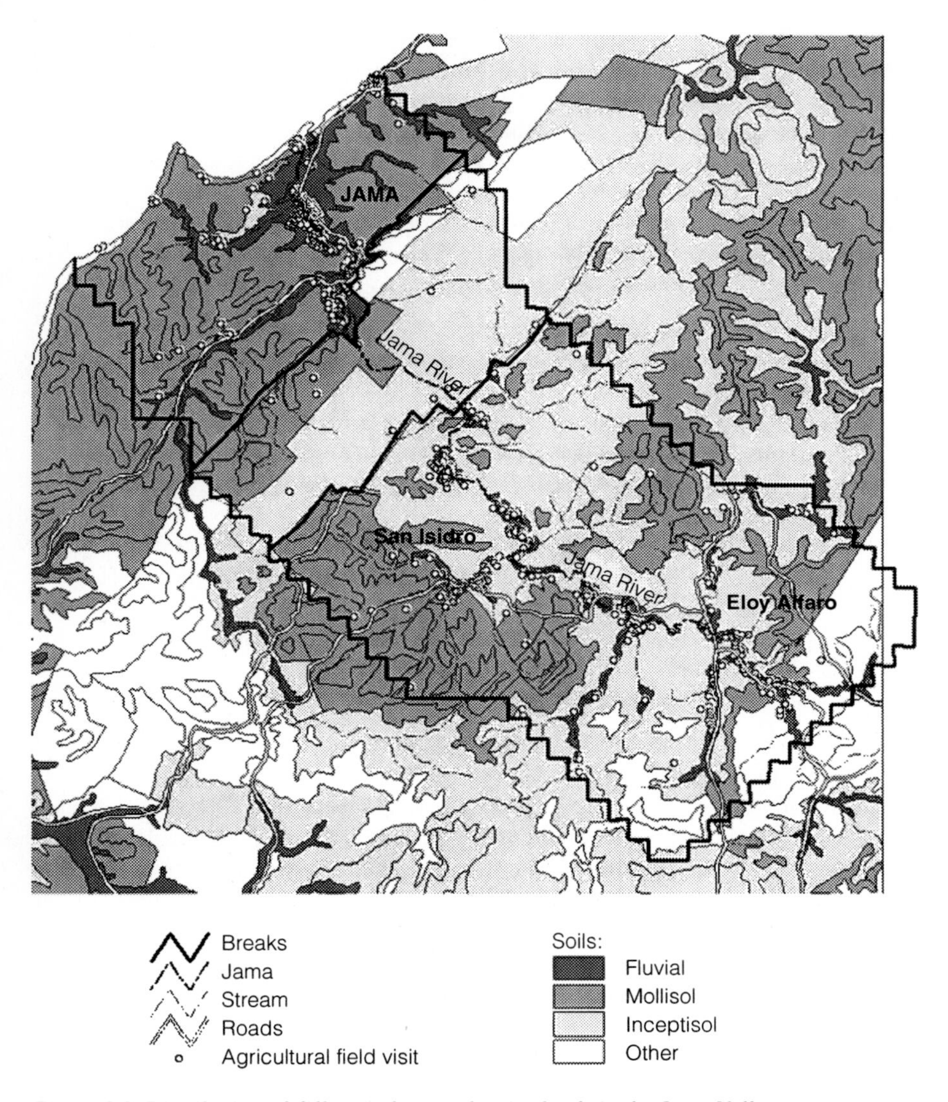

Figure 8.9 Distribution of different classes of maize lands in the Jama Valley: A working model

ized by the least productive soils for maize (no alluvium and mostly inceptisols), while the lower valley (Stratum I and adjoining western Stratum II) is dominated by productive alluvial soils and mollisols. Although lower valley mollisols are drier than those in the rest of the valley, *t*-tests comparing cob length, potential yield, and recall yield between moister and drier mollisols produced no consistent patterning; I therefore consider them as equally productive for maize (except for higher-elevation mollisols, as discussed below).

The middle valley presents the most complex picture for maize productivity. A division of soils follows the fault line paralleling the Estero Capa Perro near San Isidro. Better maize-producing soils dominate the higher-elevation areas to the southwest, south, and southeast of San Isidro. As the land-use study showed, these are the only nonalluvial lands in plantation agriculture today, which supports an interpretation of greater potential productivity for this area. The areas to the north and northeast of San Isidro, and around Eloy Alfaro, are largely inceptisols, poorer for maize production, with interspersed areas of mollisols. Rich alluvial soils form discontinuous pockets along the Jama River and its major tributaries.

If maize productivity were the most important factor driving selection of locations for habitation sites, especially early in the settlement history of the valley, then I would expect the following settlement patterns in the site survey data:

1. Alluvial lands of the lower and middle valley settled earliest, and equally early: from a point of first entry, settlements spread out rapidly from alluvial pocket to alluvial pocket
2. Upland areas of the middle valley settled next, with settlement densest in the higher-elevation areas to the southwest, south, and southeast of San Isidro
3. Subsequent spread of settlements into other areas of productive upland soils in the middle valley
4. Settlements in the hills of the lower valley would appear first in areas of good soils located adjacent to densely settled river alluvium, and spread out from there
5. The zone of inceptisols to the north and northeast of San Isidro and in the Eloy Alfaro area would be settled late in the sequence
6. The Jama Narrows area would be settled last, and lightly

If these predictions are not upheld, it may be in part because yuca served as an "equalizer" for the unequal distributions of good soils in the valley. Recall that yuca yields are not higher on alluvial soils or at higher elevation, in *brisa*-enhanced settings, while maize is much more sensitive to soil quality and available moisture, especially late in the growing season. I would therefore expect maize to dominate plantings at higher elevations and on alluvium, with more yuca interplanted, or dominating, in fields in lower-elevation mollisols, and on inceptisols. While maize yields are not reduced by wider spacing to interplant yuca, in the best maize lands it would make more sense to interplant beans, squashes, and perhaps tubers requiring higher moisture than yuca, such as *Calathea, Maranta,* and *Canna.*

Yuca develops equally well on less fertile inceptisols and on better nonalluvial soils, so if selection of locations for habitations does *not* follow predictions 2 and 3 above (i.e., settlements are not founded on better upland soils first), it may be because yuca is an important enough component in the agricultural system that potential maize yields are less of a deciding factor in settlement location. If we were to see a shift of settlement and agricultural land use to better nonalluvial maize lands later in time, this might signal an increased emphasis on maize cultivation for ease of storage (e.g., for trade or tribute), for its protein contribution to diet, or as a response to tephra fall (as discussed in Chapter 9), rather than for its productivity relative to yuca or other root crops per se. Note that there are large expanses of good nonalluvial soils in the lower valley and that such good nonalluvial soils are patchy in the middle valley. Large expanses of good maize lands might fuel a late growth of centers on the coast, especially if access to the sea became more important (e.g., for trade).

It is premature to talk about which of these scenarios is most likely, since we are not yet ready to analyze the settlement pattern data generated from pedestrian survey in terms of land productivity. However, as discussed in the 1994 project volume (Zeidler and Pearsall, 1994b), alluvial lands *were* settled first, during the Piquigua and Tabuchila phases, in both the lower and middle valley. And the first site to become a ceremonial center is San Isidro in the middle valley, located in one of the largest expanses of alluvium and adjacent to an extensive expanse of productive mollisols.

9/Staying in Balance in the Jama River Valley

In this chapter I return to a central goal of paleoethnobotany, to understand how parts of a prehistoric ecosystem functioned, and how a particular human population stayed in balance with the natural world (or what the consequences were of not achieving balance). As in all archaeological inquiry, we approach this topic from the remains of these interactions: the archaeological sites and their locations on the landscape, the record of vegetation and its change or stability over time, and the remains of foods and other culturally used plants.

In the sections that follow, I will first review what is known about the vegetation of the Jama Valley during prehistory. This will set the stage for considering the impact of humans on the environment of the valley, and the response of populations to changing conditions, both human-induced and natural. I will end the chapter by discussing people-landscape interactions during the Jama-Coaque II period, incorporating what we have learned about the plants used and the productivity of the valley, and I will speculate a bit on the stability and sustainability of this relationship. The latter topic will be explored further in Chapter 10.

THE CHANGING ENVIRONMENT OF THE JAMA RIVER VALLEY

My focus in the preceding chapters was on presenting and interpreting the record of archaeological plant remains from sites in the Jama River Valley, specifically those from the early Jama-Coaque II period. Many of the macroremains preserved by accidental or deliberate charring (i.e., burning of refuse) represent food plants or the remains of other useful plant products such as cotton and gourd. These remains, and phytoliths deposited from the decay or burning of plants brought to the sites for human use, form the basis of our understanding of Jama-Coaque subsistence.

The phytolith records from most sites tested during the project are made up not only of useful plants, but also of phytoliths deposited by floods of the Jama River and its major tributaries. These background phytolith assemblages give us some insight into the character of regional vegetation. Other sources of information on vegetation are the species of wood introduced into the sites for fuel. If people collected firewood from fields cut for agriculture, the kinds of wood preserved as charcoal are an indirect reflection of what trees were common around sites.

Because phytolith samples taken from features, floors, or trash deposits at archaeological sites represent a mixture of phytoliths deposited through human and natural agencies, and because people may select some kinds of wood for fuel while leaving others in the field, researchers prefer to investigate past vegetation using data that are not directly affected by human behavior. Naturally accumulating botanical assemblages, located away from archaeological sites, are one source of such data. One such sequence, the Río Grande phytolith profile, which was studied by Cesar Veintimilla (1998, 2000) as part of the Jama project, provides us with considerable insight into the changing nature of regional vegetation over some 4,000 years.

Naturally accumulating assemblages can also sometimes be securely identified within archaeological sites. Zooarchaeologist Peter Stahl (2000) of Binghamton University studied the animal remains from the Pechichal "Big Pit" and identified a suite of small animals that accumulated in this natural trap over the short period of its fill. This assemblage of microfauna provides insight into local vegetation in the site area.

The Pechichal microfaunal data and the Río Grande phytolith profile document vegetation at different temporal and geographical (local, regional) scales. Each contributes to our understanding of the past vegetation of the valley and provides a starting point for considering what impact people had on the Jama Valley, and in turn how changing environmental conditions affected people living in the valley.

The Río Grande Profile

The Río Grande profile is one of three terrace profiles in the San Isidro area described by geomorphologists Jack Donahue and William Harbert (1994). The profiles were cut into sediments exposed by the degradation (incision into the old floodplain) of the Jama River. Deep alluvial deposits along the main channel of the Jama River and its major tributaries in the western part of Stratum III became prime localities for prehistoric (and modern) human occupation and agriculture. Preserved as remnant terraces today, these deposits measure some 7–8 m in height above the current river channel in the vicinity of San Isidro.

Terrace sediments consist of fluvial deposits, buried A soil horizons (paleosols that represent periods of stability), and layers of volcanic ash, or tephra, that include redeposited layers (mixture of ash washed from slopes and silty clay soil). The Río Grande profile consists of nine sediment horizons, for a total depth of 8.83 m, and includes three tephra layers and two soil horizons. Based on correlation of the tephras to the dated, master sequence at San Isidro, the Río Grande profile can be divided into the chronological units shown in Table 9.1.

Ages for Deposits 1 and 2 (fluvial deposits laid down prior to the Tephra I ash fall at 1955 B.C.) were estimated by Veintimilla (1998, 2000) based on the thickness of the deposits and sedimentation rates estimated by Donahue and Harbert (1994). The upper section of Deposit 2 can be correlated to Early Piquigua, the first documented occupation phase of the valley. It is important to note that only deposits located above Tephra III contained cultural materials; Deposits 1–5 were culturally sterile, and represent fluvial deposition (Deposits 1–4) or fluvial deposits capped

TABLE 9.1 CHRONOLOGICAL UNITS FOR THE RÍO GRANDE PROFILE, AND NUMBERS OF PHYTOLITH SAMPLES ANALYZED*

Deposit 8	(1 sample)	Bioturbated deposit intrusive to Deposit 6
Deposit 7	(2 samples)	Historic and modern occupation
Deposit 6	(2 samples)	Muchique 2 and after
Tephra III		ca. A.D. 400
Deposit 5	(1 sample)	Muchique 1
Deposit 4	(1 sample)	Muchique 1
Tephra II		ca. 350 B.C.
Deposit 3	(1 sample)	Late Piquigua through Tabuchila
Tephra I		1955 B.C.
Deposit 2	(3 samples)	ca. 2525–1955 B.C. (early Piquigua)
Deposit 1	(4 samples)	ca. 2859–2525 B.C.

*Adapted from Fig. 7.15 in *Analysis of Past Vegetation in the Jama River Valley, Manabí Province, Ecuador*, by Cesar Ivan Veintimilla B., 1998. Master's Thesis, University of Missouri. Reproduced with permission.

with a buried B soil horizon (Deposit 5). Deposit 6, while containing some pottery, is also fluvial in origin, with an A soil horizon cap. Thus the major agency of phytolith deposition in the profile can be assumed to be runoff of surface soil, and the phytoliths contained within it, from the watershed of the Jama River upstream of the sampling locality.

The Río Grande profile documents profound vegetational changes in the middle valley during its prehistoric occupation (Figure 9.1). From before the documented presence of human settlement in the valley, and into the early Piquigua occupation, forest dominated the valley (60–70% of phytoliths, based on absolute counts, in each sample beneath Tephra I). Wet and open-area taxa were equally represented in Deposit 1 samples, while Deposit 2 showed an increase in open-area plants.

Following the ash fall of Tephra I, occupation by Piquigua populations continued for a brief time (late Piquigua phase), and then the valley was abandoned. Following a hiatus of 475–560 years, the Jama Valley was reoccupied by people producing Tabuchila pottery. This period of time, late Piquigua through Tabuchila, is encompassed in Deposit 3 of the profile. Forest indicators drop to just over 40%, while open-area indicators remain at levels observed in Deposit 2.

The valley was abandoned immediately after the Tephra II ash fall that ended the Tabuchila cultural phase. The hiatus lasted about 250 years. Deposits 4 and 5 in the profile are contemporary with the Muchique 1 phase (Jama-Coaque I period) reoccupation of the valley. During this time, arboreal phytoliths are documented at the lowest percentage for the profile (30–35%), and open-area taxa come to dominate phytolith assemblages for the first time.

A thick layer of volcanic ash (Tephra III) blanketed the valley around A.D. 400, but this time did not lead to its abandonment. Occupation continued with the Jama-Coaque II cultural tradition (Muchique 2 phase and after). Two samples from the Tephra III deposits were analyzed by Veintimilla (1998, 2000). While the percentages shown in Figure 9.1 indicate that most phytoliths deposited were arboreal, it is unlikely that forest regrew suddenly in response to the ash fall. Very few phytoliths were contained in these samples, due to the very rapid sedimentation rate of this deposit (e.g., there were an estimated 9,300 arboreal phytoliths in the 7-5a Tephra sample, compared to an estimated 300,000 in the 7-1 Muchique 1 sample). This is too great for a meaningful comparison of percentages.

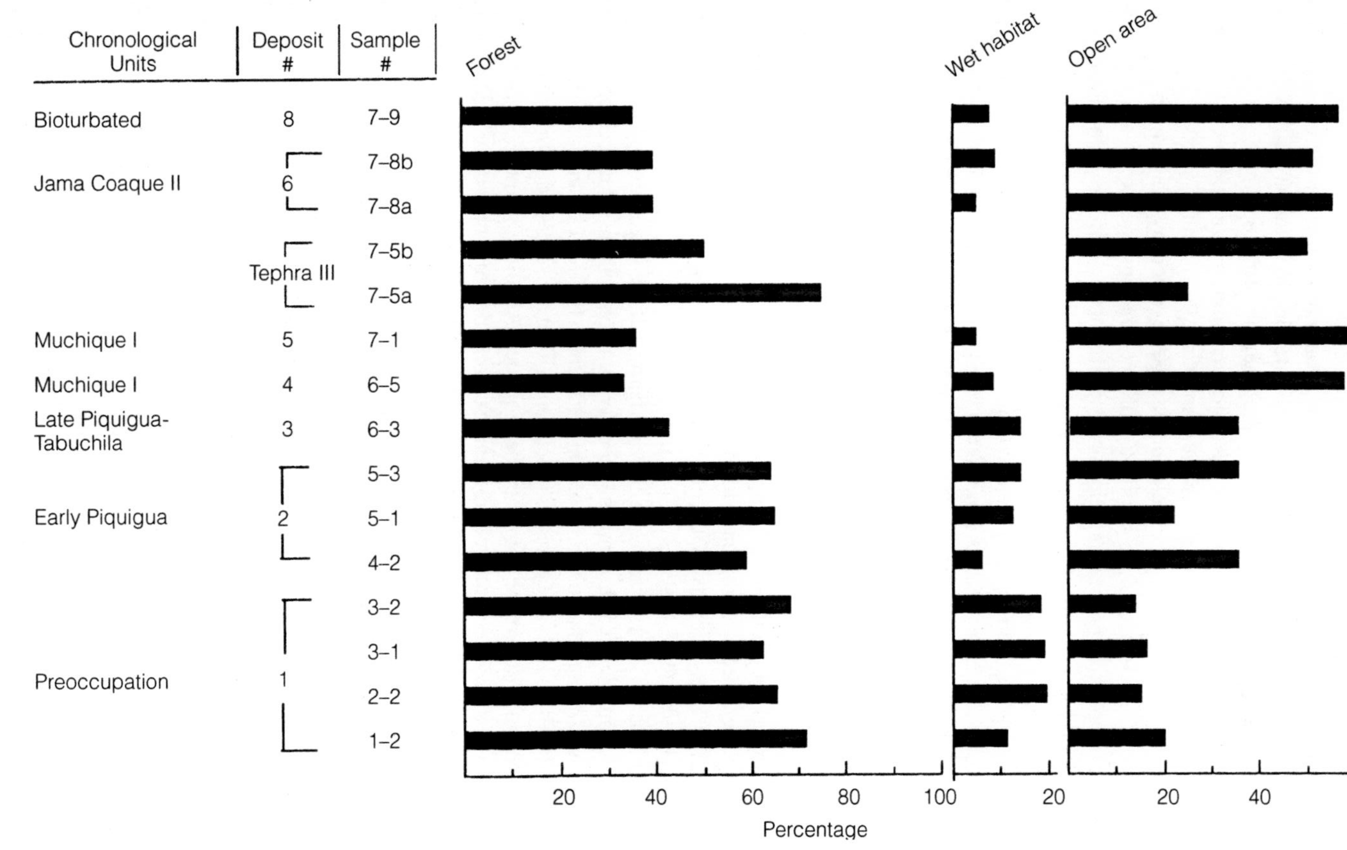

Figure 9.1 Summary phytolith data from the Río Grande profile. "Other" phytolith types are omitted, so the percentages do not sum to 100%. (Adapted from Veintimilla, 1998.)

Note that there is a slight increase in forest taxa in the two Río Grande samples contemporaneous with Jama-Coaque II. This is largely due to an increase in palm phytoliths. This is a reversal of a trend of steady decline in palm phytoliths throughout the profile, to a low in Muchique 1. Asteraceae phytoliths occur for the first time in the Jama-Coaque II assemblage, with *Heliconia*. A number of dry forest taxa present earlier in the deposits disappear in the Jama-Coaque II strata.

I suggest that the pattern of vegetation change revealed by Veintimilla's (1998, 2000) study of the Río Grande profile documents the emergence of an agroecology—a landscape managed by people for crop production—in the Jama River valley over the course of some 4,000 years. The first indication of human disturbance of the landscape, an increase in open-area plants in the context of minimal forest clearance, is seen in Deposit 2, the early Piquigua phase. During late Piquigua and Tabuchila times, there was an acceleration in forest clearance documented by a drop in arboreal indicators. This pattern continues during the Muchique 1 reoccupation of the valley after the Tephra II event: Arboreal phytoliths are documented at the lowest percentage for the profile (30–35%). Further, the abundance of open-area plants in the phytolith record suggests that tracts of the valley were maintained in increasingly open conditions. This pattern continues and intensifies in the subsequent Muchique 2 phase. Palms increase in abundance after the Tephra III event, suggesting that these useful plants were being actively managed, while other arboreal components of the forest continued to decline. This is further support for the maintenance of open habitats, with minimal regrowth of secondary forest, a pattern first seen in Muchique 1 times.

At this point, it is important to consider whether all the vegetation changes documented in the profile were human-induced. What about other influences on vegetation, such as climate change and volcanic ash fall? Below I discuss the potential effects of ash fall on the valley and its inhabitants. There is some indication that moisture levels changed during the course of the sequence. Absolute abundances of sponge spicules and diatoms (the main component of wet indicators on Figure 9.1) decline over time, with two episodes of decline and recovery during the sequence (between the top of Deposit 1 and the middle of Deposit 2, and from before and after the Tephra III ash fall). The resolution of the profile is not fine enough to identify drought events, but the coarse-grained data do indicate that moisture levels were variable.

The period of time covered by Deposits 1–6 in the Río Grande profile (ca. 2800 B.C. to A.D. 1200 and after) corresponds to a drier period documented in a number of regions in the northern neotropics (Pearsall and Jones, 2001). The wet conditions of the middle Holocene begin to taper off around 3000 B.C., and around 1000 B.C. there is a return to distinctly drier conditions. The period ca. A.D. 800–1000 sees severe and repeated droughts in some regions. The Río Grande results are thus consistent with a regional drying trend, but there is no evidence that this dramatically affected vegetation in the valley: Dry tropical forest (albeit in remnant form) remained the climax vegetation of the valley throughout this period. Another factor contributing to decreasing moisture was reduction in tree cover resulting from forest clearance.

Perhaps the most direct supporting evidence that humans were responsible for converting the Jama Valley from a dry tropical forest to a mosaic of open habitats and remnant trees comes from the archaeological record. Here I am speaking of the many cultivated plants recovered from sites tested during the project. Appropriate habitats for these sun-loving plants had to be created. Maize phytoliths appear for the first time in the Río Grande profile in Deposit 2, in association with *laurel* (*Cordia*), an open-area colonizer species. Marantaceae and Cannaceae phytoliths are present in Deposit 1 and after.

Pechichal Microfauna

Pits like Feature 5 at the Pechichal site are natural traps for small animals: Food refuse attracts rodents and other scavengers, but the steep sides of the pit hinder escape (Figure 7.1). A total of 12,591 bone fragments were recovered from the Big Pit and analyzed by Peter Stahl (2000). Specimen fragmentation was severe: Only 3,853 vertebrate remains could be identified minimally to class. The archaeological faunas appear stratigraphically distributed in two distinct groups: (1) concentrations of taxa characterized by many specimens, from many individuals, and (2) dispersed examples of taxa characterized by a few specimens and few individuals. The stratigraphically concentrated faunas also tend to be more complete (i.e., with more different elements), and the dispersed faunas less so.

Some of the stratigraphically concentrated faunas likely represent food remains; for example, concentrations of burned remains of aquatic faunas (marine shell, river shrimp, fish) and armadillo were found in the lowest units of the pit. Many others represent episodes of natural entrapment, however. Among the small animals recovered in relatively complete form were frog or toad, small snakes, and small (most likely rice rat) and medium (most likely spiny rat) rodents (Figure 9.2). The stratigraphically dispersed faunas, by contrast, were by and large potentially important food animals like common opossum, howler monkey, capuchin monkey, agouti, rabbit, peccary, and cervids (deer). Larger birds, such as tinamou and heron, also fall into this class.

When the natural histories of the animals recovered from the Big Pit are considered along with their representation in the pit, valuable information about the local ecology is revealed. The faunal data indicate that during the time the pit was open (i.e., during the Muchique 2 phase) it was located close to the modified edge of a forest fragment. Stahl (2000) argues, for example, that among the entrapped fauna are an abundance of hardy generalists, animals adapted to forest edges, who readily invade forest fragments. Small rodents and snakes proliferate in secondary forests, agricultural fields, and high-diversity forest edges in the tropics. Rice rats, the most likely taxon making up the small rodent group, are at home in secondary growth, particularly growth around agricultural fields and villages. Rodents that favor drier, open conditions—including cotton rats, grass mice, and cane mice—were also found. The stratigraphically concentrated fauna suggest, then, that Pechichal was surrounded by a mosaic of open habitats, fallow fields, and secondary growth that resulted in forest fragmentation.

The potentially important food animals recovered from the Big Pit would also be at home in such an environment. Fallow fields attract peccary, whitetail deer, agouti, and rabbits. Opossum and armadillo thrive in secondary growth surrounding forest fragments, as do tinamous and doves, and Howler monkeys.

Studies by Veintimilla (1998, 2000) of a naturally accumulating phytolith record (the Río Grande profile) and by Stahl (2000) of the faunas of Feature 5 of the Pechichal site provide us with considerable insight into past environmental conditions in the Jama River Valley. The local picture of modified forest fragments that

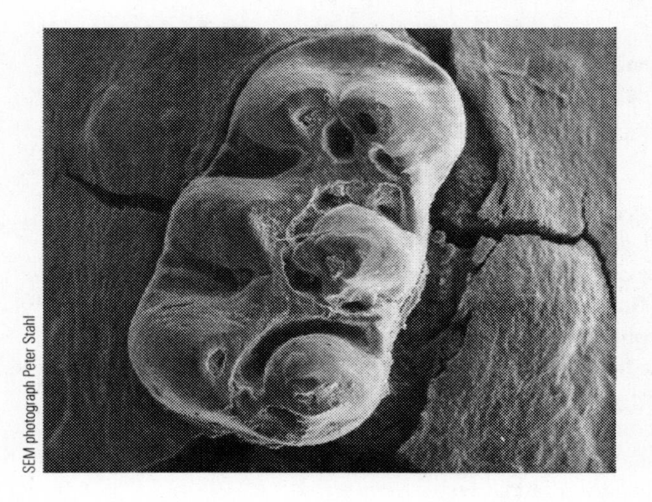
SEM photograph Peter Stahl

*Figure 9.2 Presence of microfaunal remains like this rice rat (*Oryzomys*) molar, recovered by flotation, provides valuable clues about local environments. Rice rats are generalist rodents.*

emerges from the Pechichal faunas corresponds well to the regional view of the valley in Muchique 2 times. It is against this backdrop of extensive and sustained human modifications of the environment that I turn to the periodic disasters that beset the inhabitants of northern Manabí: volcanic ash fall.

DISASTERS IN PARADISE

Tephra Falls in the Jama River Valley

Three tephra layers, that is, layers of volcanic ash (and sometimes larger solid matter) ejected from volcanoes that falls to the ground from eruption columns, have been identified and sampled throughout the Jama Valley. By means of radiocarbon dates taken immediately above and below the tephras, and analysis of ceramics and site stratigraphy, Zeidler (1994a, b) has established the chronology of these ash fall events. The source of one ash fall has been determined: The eruption of Pululahua volcano, located on the equator outside Quito, was responsible for the Tephra II event (Isaacson, 1994). Isaacson and Zeidler (1999) have demonstrated that the effects of the Pululahua and other ash falls were felt over a wide area of western Ecuador, in significant portions of an area lying between 0°30' north latitude and 1°30' south latitude. This includes present-day northern, central, and part of southern Manabí province, and the adjoining montaña and highland zones to the east.

A number of studies have been done of the effects of volcanic ash fall on vegetation. One case useful for understanding potential effects in the Jama Valley is the 1943–1945 activity of Paricutín volcano in the state of Michoacán, Mexico, studied by Eggler (1948, 1963), and later by Rees (1979). Before volcanic activity began, 75% of the land in the Paricutín region was pine-oak forest, with maize agriculture confined to flat and gently sloping lands. Hill slopes were primarily forested. Rainfall was highly seasonal, and caused rapid soil erosion on deforested slopes.

Study of the area in 1945, following over two years of active lava flow and ash fall, revealed that ash eroded very rapidly on hill slopes: A third of the ash deposited on steep slopes had already eroded into the valleys. The reduction in the depth of ash on slopes occurred even as ash was still falling. Erosion during the rainy season was the main avenue for redeposition of ash; during heavy rains, ash flowed as sheets of mud. By 1959, even in areas where all vegetation was buried and killed by ash fall greater than 2 m in depth, once the ash eroded away, trees, shrubs, and grasses began to regenerate from roots and plant bases. In areas of ash fall of about 50 cm or less, buried plants regenerated through the ash in a few years; however, Rees (1979), in his study of the area 20 years after the eruption, found that many trees and shrubs that had initially survived later succumbed. New plants would not establish on primary ash deposits unless organic matter or soil was mixed in. Seedling germination was apparently inhibited by chemicals in as little as 10 cm of ash. Maize seeds planted in ash did grow, however, as long as plant roots were able to reach the old ground surface (i.e., in ash up to 30 cm deep).

Lack of nutrients in tephra deposits is the most important factor limiting the reestablishment of agriculture after a volcanic disaster (Thornton, 2000). Nitrogen and phosphorus are both very low—too low for maize to grow without massive additions of fertilizer. Even when soil and ash are mixed, maize yields will be low until humus builds up in the soil. On a small scale, crops can be planted in pits dug through the ash to the old soil surface. Being labor-intensive, however, this solution is not practical except for house garden plots. Field production on flat terrain immediately after ash fall is impossible unless ash is thin enough that it can be turned into the old soil by plowing, or is extensively mixed with soil through redeposition. Otherwise, farmers must wait for natural vegetation succession to occur. Twenty years after the Paricutín volcanic eruptions, for example, succession on thick ash (> 50 cm) had not yet reached the point where cropping was possible, and only a few fields with 25–50 cm of ash were back in production.

Ash from Paricutín volcano killed all above-ground portions of plants when it accumulated to a minimum depth of 0.7–2.0 m. Small trees were most affected, dying in accumulations of only 20–30 cm of ash. Large trees were killed by deposits of 30–60 cm, primarily because branches broke under the weight of the ash. Medium trees fared best, surviving ash accumulation of up to 2 m because their branches were more flexible. Breakage and starvation (from ash either covering leaves or causing leaf fall) had greater impact on tree mortality than depth of ash around the base. If ash was carried away by rain as fast as it fell, trees recovered quickly.

It is difficult to generalize about the fertility of soils derived from tephra, since soil fertility is determined by the composition of the parent material, the weathering regime, the age of the soils, and water retention and permeability (Ugolini and Zasoski, 1979). Young soils, for instance, vary in fertility according to organic matter content, and nitrogen is typically a limiting factor. In the Jama River Valley, young alluvial soils admixed with tephra and organic matter that are well drained are among the most desirable soils for commercial agriculture today. And as the soil analysis showed, both alluvial and nonalluvial soils selected by farmers for maize cultivation are quite fertile.

In addition to the tephra layers identifiable in profiles from archaeological sites in alluvial settings and in river cuts, thick layers of redeposited ash and soil, eroded from hillsides and carried downstream by the Jama River and its tributaries, are prominent features of many terrace profiles (see Figure 4.2). Redeposited ash layers are often much thicker than pure tephra. Based on the sources I consulted, erosion of tephra from the steep terrain of the uplands of the Jama River Valley should have been rapid. Erosion would have been especially pronounced in cleared upland areas, and during the rainy season.

Even as little as 20–30 cm of ash would have affected small trees, shrubs, and herbaceous vegetation throughout the Jama River Valley, at least to the extent of killing them back to the ground. Regeneration of perennial trees and shrubs would have been rapid, with perhaps slower reappearance of annuals until original ground surfaces were exposed, or soil or organic matter was mixed with the ash. Erosion carried ash and soil into the river and stream valleys, creating deposits deep enough to kill any vegetation that survived the initial ash fall, and to make regeneration of perennials through the ash difficult. In areas where quantities of soil and organic material were mixed with ash, however, plants may have become quickly reestablished from seed.

Profiles for archaeological sites tested during the Jama project provide some insight into the depth of tephra deposition in the valley bottomlands (Table 9.2). The thicknesses in Table 9.2 should be considered minimal estimates; in most cases reworked tephra deposits (i.e., ash plus soil eroded from slopes and upstream) occur above the pure tephra deposits and profile compaction has occurred. Total thickness of tephra and reworked tephra is variable, but Donahue and Harbert (1994) note that deposits of 100–200 cm are not uncommon in the middle valley.

Looking just at the thickness of deposits designated as tephra on profiles, however, is informative. Note, for example, how tephra thickness varies even within the relatively restricted geographic area around San Isidro (San Isidro profiles, plus Río Grande and Capa Perro): For Tephra III, a range of trace to 167 cm of tephra is documented. Pechichal, located downstream from this group of sites, has 130 cm of Tephra III in the vicinity of the Feature 5 pit. El Tape, located closest to the coast (and furthest from the source of the ash fall), has only a thin Tephra III layer. Distance from the source is one factor that influences tephra thickness, but very localized factors must also be at work. In the case of deposition at San Isidro, some areas may have been protected from prevailing winds by hills. The thickness of reworked tephra in alluvial deposits is also influenced by local topographic features such as the steepness of slopes above the valley bottom, as well as drainage patterns, the force of river flow, and the distance of the test locality from the river channel.

Tephra II, which ended the Tabuchila occupation of the valley, is documented in fewer profiles than Tephra III, the Jama-Coaque tephra, but shows variation in thickness similar to that from the later event. Tephra II, identified as the massive Pululahua eruption (among the strongest documented for the Holocene), resulted in an occupation hiatus of about 250 years. Tephra I, which ended the Piquigua occupation, appears to have been the thinnest of the three tephras, but resulted in the longest hiatus, 475 years or more. Thickness of tephra fall alone thus does not account for the different responses of the cultures occupying the valley (i.e., abandonment after Tephra I and Tephra II, and continuity after Tephra III). This point is vividly illustrated by comparing Tephra III thicknesses at Pechichal, in Stratum III, and El Tape, in Stratum I. In both cases, populations producing Muchique 2 ceramics reoccupied riverine village sites after redeposition of tephra ceased, and dug features that penetrated into the primary Tephra III deposits. That there was 130 cm of tephra at Pechichal and 7 cm at El Tape seems to have made no difference: Occupation at both villages resumed after the tephra event, with no indication of a lengthy hiatus. Why was this?

TABLE 9.2 THICKNESS OF TEPHRA DEPOSITS (CM)*

Site	Tephra I	Tephra II	Tephra III
San Isidro sectors			
XII/C11	43	17	20
XVIII/A1	—	73	7
V/A1	—	33	7
V/B1	46	4	38
XXX/A1	—	42	trace
Río Grande	5	5	178
Capa Perro	—	123	—
Pechichal	—	—	130
El Tape	—	—	7

*Estimated from site profiles in Donahue and Harbert (1994), Zeidler (1994a, b), and unpublished project documents.

CHANGING PEOPLE-LANDSCAPE INTERRELATIONSHIPS

As Isaacson, Zeidler, and I have each discussed (Isaacson, 1994; Isaacson and Zeidler, 1999; Pearsall, 1996), the different responses of Piquigua, Tabuchila, and Jama-Coaque populations to the natural disasters of volcanic ash fall, vegetation die-off, crop destruction, and massive erosion have their roots in a number of differences among these cultures. My focus here is on what was different about the relationship of Jama-Coaque populations to their environment. Three factors come to mind.

First, agriculture in the valley as a whole must have recovered quickly from the Tephra III event to sustain the dense populations documented by the pedestrian site survey, and this implies that the uplands (nonalluvial lands) were an important part of the agricultural system at the time of the disaster. The inhabitants of the Jama River Valley did not initiate agriculture in the uplands in response to destruction of floodplain-based agriculture: Forest clearance was well underway prior to this time. In fact, many upland fields already cleared for agriculture by Muchique 1 farmers would have recovered almost immediately from the Tephra III fall, at least with the onset of the next rainy season, and with newly deforested upland fields would have produced food sufficient to feed the valley.

Second, populations occupying valley bottom sites must not have been solely dependent for their sustenance on cultivating the alluvial lands around them, but must also have had access to agricultural products and/or fields in the uplands. Valley bottom villages, including ceremonial centers, were quickly reoccupied, but alluvial lands would not have been productive for some time. Since the uplands were occupied during this time, albeit not as densely as in subsequent periods, this access was not because there were expanses of unused lands, but more likely because the inhabitants relied on social mechanisms. While kinship ties to villages in the uplands may have provided access to these lands, formal redistribution networks managed from the ceremonial centers were very likely of critical importance for provisioning the populations of valley centers.

Third, agriculture may have changed in composition by Jama-Coaque times, in comparison to the earlier Formative period pattern, in ways that facilitated recovery from agricultural disaster. Not only had agricultural practices diversified by Muchique 1 times to include cultivation in the uplands, but there is also some evidence that a shift in the relative importance of maize and root/tuber foods accompanied this diversification. While it is impossible to compare Formative period subsistence to subsistence during Jama-Coaque I and II directly due to preservation factors, comparisons within the Muchique sequence give some insight (Figure 9.3). As the graph shows, maize apparently increases in abundance relative to other food plants (tree fruits, beans, root/tuber foods) between Muchique 1 (pre–Tephra III) and early Muchique 2 (immediately post–Tephra III). The ratio of maize to other foods doubles, and maize abundance per 10 liters of soil increases nearly 8-fold, while other foods show only a 4-fold increase. In the latter part of the Muchique 2 phase, maize abundance declines. It rises again following Muchique 2, but not to the levels seen previously.

The relationship of these changes in food abundances to the Tephra III event is suggestive. Of the available crops, perhaps maize was best adapted to the altered growing conditions faced by Muchique 2 populations after the ash fall disaster. Annuals, like maize, yield after a single rainy season. Maize can be consumed as green corn even before the crop is fully mature, while long-growth root crops would not be available as quickly. As shown in Chapter 8, maize yields well in fertile upland soils, and such soils are available in both the lower and middle valley. Arboreal resources would be the hardest hit by the Tephra III event—mature, producing fruit trees would have been decimated by ash fall, and replacements grown from roots or by reseeding would have taken years to resume production. The best orchard lands may have been out of production for a generation. The eventual reestablishment of a mixed cropping strategy of root/tuber foods, maize, and arboreal resources, as suggested by the late Muchique 2 and later data, may have contributed to long-term stability of subsistence in the valley, however. As other researchers, including David Rindos (1984), have noted, seed crop agriculture tends to be less stable than root crop or mixed subsistence systems. I will return to the topic of the stability of tropical forest agriculture in Chapter 10.

There are also some hints of crop specialization in sectors of the valley by the time of the Tephra III disaster, which would have further mitigated the effects of the tephra fall on food production. If populations living in the ceremonial centers in the valley bottom specialized in cotton production, and perhaps foods and plants important in ritual and feasting, while populations living in the uplands produced food staples for provision to elites, then disruption of bottomland cultivation would be less devastating for the survival of populations.

In his summary of recent volcanism in protoclassic and later Mayan sites in Costa Rica and El Salvador, Payson Sheets (1983, 1994a, b) emphasized the importance of considering the size and nature of the sociopolitical unit, as well as the magnitude of the event, for understanding the impact of volcanism on societies. In the case of three tephra falls that affected only small segments of ongoing Mayan societies, recovery was facilitated by exchange with nondevastated adjoining areas. In contrast, the more extensive Ilopango fall wreaked havoc on numerous units of the society, and affected many thousands of square kilometers: The scale of the disaster exceeded the boundaries of the society. This made recovery through local efforts impossible, and resulted in abandonment. While there are relatively few comparative cases to draw from, the latter scenario seems to be the exception:

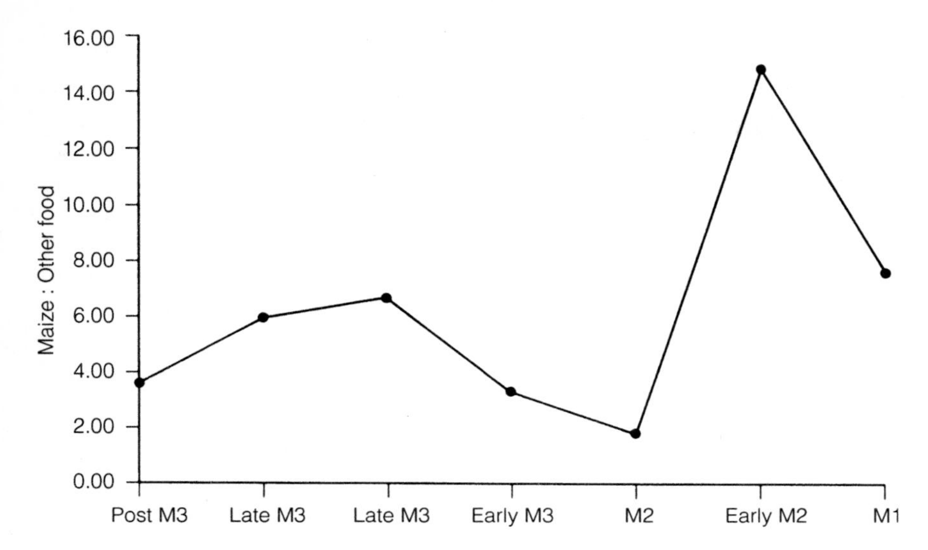

Figure 9.3 Ratio of maize to other food plants over time in the Jama Valley. The cultural sequence runs from early (right) to late (left). M1: Muchique 1; M2: Muchique 2; M3: Muchique 3. From "A Case Study from the Jama River Valley, Manabí, Ecuador" by Deborah M. Pearsall, in Case Studies in Environmental Archaeology, *ed. E. J. Reitz et al., 1996, Kluwer Academic/Plenum Publisher. Reproduced with permission.*

> What is striking to me in attempting to see some sort of patterning in these cases is the remarkable resiliency, in the long run, of human societies in dealing with volcanism. Very few eruptions caused significant cultural-demographic effects that lasted long enough to leave marked effects in the archeological record. (Sheets 1983:291)

This was especially the case if groups did not rely on "built" environments for agriculture (i.e., terraces, raised fields, irrigation systems) and had a diversified subsistence base. In the Jama case, while analysis of ceramics suggests there was cultural continuity before and after the Tephra III event, there was a shift in the complexity of political organization. In other words, Muchique 2 ceramic tradition marks the beginning of Jama-Coaque II, the Integration period culture. Did the cultural response to the natural disaster stimulate these changes?

It is safe to assume that crops on alluvial lands would have been destroyed and the lands rendered unusable by tephra deposition of the magnitude documented in Jama profiles. How long alluvial lands, the most fertile in the valley, would have been out of production would have depended on the local depth of the tephra, and how much soil and organic matter was mixed with it. Upland plantings also would have been destroyed by ash fall, but once the ash was washed away, productivity would rapidly recover. Forested uplands also would have recovered quickly, as shrubs and trees regenerated from roots. Increased clearing of uplands for planting, to replace lost alluvial lands, and a focus on annual crops over perennial species would have been logical short-term responses to deep tephra deposits in the valley.

• • •

PART III: SUMMARY

An agricultural system based on use of productive alluvial lands may have ill-prepared the first inhabitants of the Jama River Valley, the Piquigua and Tabuchila populations, to cope with tephra fall disasters, while a system incorporating both alluvial lands and upland plots may have facilitated the survival of Jama-Coaque II populations. Is this sequence, alluvial-based agriculture followed by the onset of shifting cultivation in uplands, a typical example of agricultural evolution in the lowland tropics? In a word, no: Most models posit the appearance of swiddening *before* use of annually flooded lands. Let us explore this topic further as we look beyond the Jama River Valley.

10/The Evolution of Tropical Forest Agriculture

> These long-vanished riverine people, who were the most numerous inhabitants of the Amazon Basin around A.D. 1500, have the best claim to be regarded as typical representatives of Tropical Forest Culture; but it was the riverine peoples who were first destroyed by the combined effects of European diseases, missionization, and the slave trade.
>
> Lathrap (1970, 47)

Archaeologist Donald W. Lathrap (1970) argued, following cultural geographer Carl Sauer (1952), that prime agricultural lands such as riverine floodplains would have been the first utilized by lowland agriculturalists, and that such lands permitted permanent settlements to be established. Shifting forest cultivation—the slash-and-burn systems common today in the Amazon basin and other tropical lowland regions—represented a necessary adaptation of marginalized populations displaced from prime lands. While other scholars may disagree with Lathrap's scenario for the emergence of agriculture, there is increasing awareness that subsistence patterns and lifeways of surviving tropical forest agriculturalists are, at least in part, a reflection of contact, the Colonial experience, and modern social and political conditions—and are, at best, a clouded window into pre-Columbian adaptations.

The model for the primacy of permanent to semipermanent agriculture in the lowlands, and the secondary development of shifting cultivation, has been well developed by William Denevan. I rely primarily on his 2001 book, *Cultivated Landscapes of Native Amazonia and the Andes*, to articulate this model, but also incorporate insights into tropical forest agriculture from a number of other sources. In this chapter, I attempt to evaluate the extent to which the findings of the Jama project, including settlement pattern data and archaeobotanical and vegetational records, support Denevan's (and Lathrap's and Sauer's) model for the evolution of tropical forest agriculture. Further, I will consider to what extent the dry forest setting of the Jama River Valley influenced the nature of agricultural evolution in the valley.

Most of the large expanses of dry forests of the neotropics were readily accessible to European settlers, and these desirable lands, as well as fertile soils of riverine settings, were converted early to uses such as plantation agriculture and cattle husbandry. There still remains much we can learn from accounts of surviving tropical forest agriculturalists, however, when this knowledge is tempered by consideration of changes brought about by contact, and placed in the context of a thorough understanding of the tropical forest environment.

I first review tropical forest agriculture in general terms, and then in its two major variants, shifting cultivation and annual cropping of floodplain habitats. For these discussions, I consulted Denevan (2001), Goldammer (1992), Kellman and Tackaberry (1997), and Moran (1993). After discussing the probable nature of dry forest agriculture, using sources on this type of forest, including Gentry (1995) and Maass (1995), I return to the Jama case, and consider what our data say about the evolution of agriculture in the lowland tropics, as well as discuss issues that remain unresolved.

WHAT IS TROPICAL FOREST AGRICULTURE?

There are two dominant processes in tropical ecosytems: a complexity of biological interactions, and a deficiency of plant mineral nutrients in the soil. Biotic complexity creates special problems when humans attempt to create simpler, artificial systems such as agricultural fields. In essence, agriculture is a new type of ecosystem for the forested tropics: low-diversity, herbaceous, nutrient-demanding crops, inefficient at competing with other plants or animal predators, are substituted for a high-diversity, well-adapted woody flora. Farmers traditionally reduce risk by planting a range of varieties of crops (local varieties, or landraces) and by selecting crops that do well under local conditions. In a sense, shifting cultivation avoids the problems of maintaining low-diversity cropping systems: Fields are "abandoned" when weed competition, pest infestations, and declining soil fertility depress productivity. Cropping on annually renewed alluvial lands provides another solution: Pests and weeds are swept away in annual floods that deposit new soil.

In tropical forest ecosystems, nutrients essential for plant growth are bound up in plant biomass; unless these are released, insufficient soil nutrients may limit crop growth. Burning of vegetation releases nonvolatile mineral nutrients (carbonates, phosphates, and silicates of nutrient cations) stored in plant tissues to the soil. These are soluble and enter the soil when the rains begin. This raises soil pH and reduces aluminum levels, creating more favorable conditions for crops. The same effect is achieved by in-situ decay of cut vegetation left as mulch. Burning also sterilizes the soil, however, killing pests and weed seeds. While alluvial soils may be highly fertile, not all are: Whether soils are high in weatherable minerals depends on the geology of the watershed. Further, higher terraces that experience less frequent flooding generally have poorer and more weathered soils than terraces closer to the river.

Today in the Jama River Valley, the steel machete is the all-purpose agricultural tool of choice (see Chapter 8). This is true throughout the tropical lowlands, beginning with the introduction of these useful tools soon after contact. But what tools were available in pre-Columbian times for creating and maintaining agricultural fields? Denevan (2001) provides the following useful summary for South America:

> *Stone axe*: a shaped, sharpened stone hafted to a wooden handle. Compared to metal axes, the cutting edge is blunt, and would mostly smash, rather than cut, wood fibers. Ethnographic accounts vary, but several days were typically needed

to fell a single tree. Stone axe use was likely supplemented by girdling and firing trees.

Macana (used with the *garabato*): a wooden tool for slashing forest undergrowth and for weeding (basically a wooden machete). The *garabato* is a forked branch used as a hook to pull vegetation toward the worker who cuts it with the macana.

Digging stick: a branchless wooden pole, sharpened and fire-hardened at one end, up to 2 m long, which was plunged into the soil to make a planting hole.

Triangular- or paddle-shaped digging tools (*lampas, spades, shovels*): used for digging, hilling, planting, weeding, and breaking up clods. These are one-piece tools made of hard wood, such as *algarrobo* or palm; in some regions, metal blades or animal scapulae were attached to wooden handles.

Mattock: a short-handled hoe, basically a hooked stick with an attached blade of wood, stone, bone, or metal (in some regions).

Agricultural fields of many forms exist in the tropical lowlands, from "ordinary" fields in which there is no obvious human modification of terrain, to artificially raised planting platforms with elaborate systems of drainage channels. Our research to date in the Jama River Valley indicates that during prehistory most fields were of the ordinary type, except for some alluvial fields with limited water control features. Ordinary fields are characterized by shifting cultivation, the rotation of a few years of cropping with a moderate to long fallow (resting) period. Shifting cultivation systems are nondestructive of environmental resources and energy-efficient if practiced at low intensity and in relatively flat terrain. Where soils are fertile, fallowing may be only for a few years (short-fallow cultivation). The best soil and good rainfall permit annual cropping or, in regions with enough rainfall, multiple crops per year.

Most tropical forest agriculturalists today rely on a limited number of staple crops for energy and protein needs. While tropical forest agriculture is often characterized as highly diverse, many crops such as fruit trees are secondary dietary items. Major tropical crops can be grouped into three categories: high-bulk starchy crops (primarily root crops), cereals, and legume seeds (pulses). High-bulk root crops, such as yuca, yield well but are relatively poor sources of protein. They also store poorly (or require substantial processing into a storable product). Grains such as maize produce less bulk food but better protein and are easier to store.

Monocultural fields are common today in the lowland tropics, even among native farmers; for instance, a field might have many different species in it, but 80% of crop biomass may be yuca. Another common pattern is zonation: sectors for one crop, or crop combination, separated from other sectors. Monocultural and zoned fields are common in the Jama River Valley today. Mixed cropping, or polyculture, is still widespread in the neotropics, however. Many combinations of annuals, perennials, herbaceous, and arboreal species have been documented ethnographically, as has the folk knowledge of what crops make good companions for others.

In Chapter 8, I described how farmers in the Jama River Valley grow maize and yuca today. What I was describing is a modified form of shifting cultivation, in which new fields are cut from forest (mostly second-growth), used for a few years (3–4 being typical), and then sown in pasture grass and effectively taken out of the agricultural cycle. In traditional shifting cultivation, fields are allowed to regrow during a fallow period, and then re-cut.

Fields in a shifting cultivation system are "abandoned" when greater yields for the same labor input can be achieved by making a new field (Figure 10.1). Potential factors affecting yields and labor input include declining soil fertility, buildup of insect pests, and invasion by weeds. Available technology (steel tools, insecticides, and the like) and labor are also factors. The role of insect pests in the decision to fallow a field is unclear; some studies indicate that pest outbreaks are episodic, rather than increasing with years a plot is used. And some crops, like yuca, are resistant to insect damage. High crop diversity in a field can reduce pest damage, but not just diversity per se—studies indicate the "right" mix of crops is needed, or the farmer may just provide pests with tasty alternative hosts. Traditional knowledge plays an important role here.

Few data support dropping soil fertility as a cause for abandonment. This is not to say that fertility does not decline over time, but rather that other factors, especially weed invasion, lead to the decision to fallow a field that still may be fertile enough to grow crops. Adding fertilizer increases yields on permanently cropped fields, and there are ethnohistoric accounts of fish or fish heads, as well as *guano*, being used as maize fertilizer by traditional agriculturalists. Periodic firing of fields sterilizes soil and releases some nutrients back into it, allowing for longer cropping cycles. Intercropping with nitrogen-fixing crops, such as beans and peanuts, can also help sustain field fertility.

The negative effect of weeds in agricultural fields is basically competitive, and as every gardener knows, once weeds, especially grasses, invade, they quickly get the upper hand. How do traditional agriculturalists keep weeds from getting established? The first thing, Jama farmers told me, is to get a good, hot burn to kill weed seeds in the soil. In established fields, two effective methods of weed control are periodic weeding, and interplanting crops to include some that develop rapidly and cover soil to shade weeds out. Spreading plants like sweet potato serve this function, but as described in Chapter 8, yuca can also be used this way in maize fields: If yuca is cut back when maize is planted, its later growth can help shade the ground around the taller maize plants. Organic mulches also provide weed control; there is a variant of shifting cultivation known as slash and mulch, in which vegetation, once cut, is left to decay in the field rather than burned (Thurston, 1997). Slash and mulch systems are most common in the wettest tropics, where it is difficult to get a good burn and there is enough mass of vegetation to get the beneficial effects. For seed crops, vegetation is dropped on top of broadcast seeds, and shoots emerge through the rotting mulch.

In his research among the Kayapó of Brazil, Posey (1984) demonstrated that fallow fields are anything but "abandoned" by forest agriculturalists. After production peaks in new fields (after 2–3 years), fields continue to bear domesticated plant products for many years: 4–5 years for most root crops, up to 25 years for *achiote* trees. Further, as forest regenerates, many useful plants appear, including foods, medicines, fish and bird baits, thatch, paints, oils, insect repellents, firewood, construction materials, fibers for ropes and cords, body cleansers, and products for craft production, among others. For the Kayapó, fallow fields are perhaps most important for concentrations of medicinal plants, a phenomenon noted in other areas of the tropical lowlands. Finally, game animals are attracted into fallow field areas. Balée (1994) has discussed similar patterns of use among the Ka'apor, and other examples could be given.

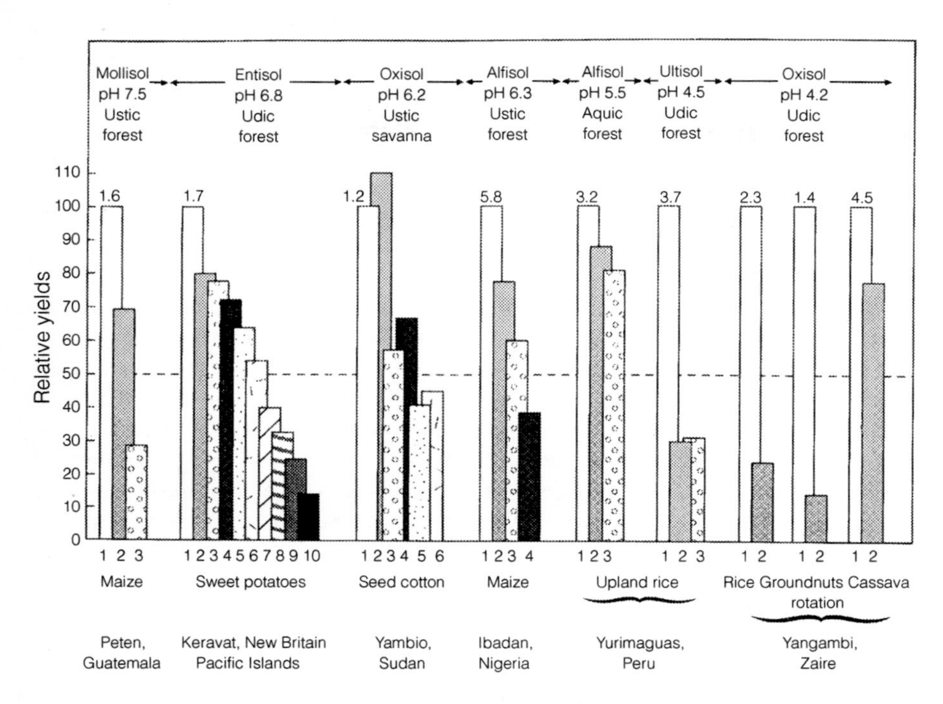

Figure 10.1 Yield decline with continued cropping in shifting cultivation systems. From Tropical Environments: The Functioning and Management of Tropical Ecosystems, *by Kellman and Tackaberry, 1997, Routledge. Reproduced with permission from Thomson Publishing Services.*

What Posey and others have documented is that fallow fields are not just large forest gaps, but exhibit a distinctive pattern of regrowth. Fields are first covered in the weedy vegetation of the last crop cycle, stump-sprouts (especially in dry forests), perennial plantings, and pioneer tree species. Over time, open-area plants are shaded out, and shorter-lived pioneer tree species are replaced by slower-growing forest trees. Fruit trees may persist or even spread. Regrown fields have a simpler plant composition than mature forest.

A useful way to view this process of forest clearance and regrowth is from the perspective of "disturbance or interference" ecology (Goldammer, 1992). From this viewpoint, human manipulation of vegetation is a natural process in the broad time scale. Certainly it has been part of the ecology of the neotropics for 12,000 years or more. If we can avoid subjective valuation of contemporary forest destruction, shifting cultivation can be understood as the type of small- to large-scale disturbances that are key processes in speciation, and play an important role in evolution. The "seral forests" created are highly valued, as are the edge habitats. As Balée's (1994) work among the Ka'apor demonstrates, these anthropogenic forests are highly diverse and are rich sources of culturally valued plants.

The process seen in the Jama River Valley today of forest being replaced by sown pasture grasses can also be an unplanned outcome of shifting cultivation. If aggressive grasses are part of the weedy vegetation and fires are frequent, savannas can become established rather than secondary forests. Fire can, in effect, create islands of forests in savanna vegetation. Although this creates new open habitats and expanded edges, it decreases the amount of land available for traditional agriculture.

This brings up the potential role of fire in intensifying shifting cultivation systems in the lowland tropics. Growing populations eventually reach a point where uncleared forestland becomes scarce. Fields must be kept in production longer and/or fallow periods shortened. If weed invasion is the primary limiting factor behind lengthening cropping cycles, can in-field fire be used to manage soil during the cultivation cycle (burning off weeds, crop residues, and forest litter), in order to extend the productive life of a field? Other effective weed control measures, such as mulching and shading, were available prehistorically, and weeding using cutting tools such as the mattock and *macana* was possible, if labor-intensive.

Kellman and Tackaberry (1997), among others, hold that continuous fallow shortening will fail to achieve a new equilibrium level of organic matter or nitrogen in tropical soils, lead to local extinctions of woody taxa, and eventually reach a point where woody fallow does not reestablish. Derived savanna and collapse of the agricultural system follow. A total of 50 people/km^2 is considered the maximum carrying capacity of shifting cultivation systems, based on constraints of cropping time and fallow periods. Essentially, then, intensification of shifting cultivation is not environmentally sustainable, that is, agriculture cannot be practiced indefinitely without ongoing deterioration of the environment—especially the soil resources—in which it exists. Intensification requires good soils, and long-term sustainability remains questionable.

Are there any avenues by which intensification of shifting cultivation could take place in the lowland tropics? Denevan (2001) and Balée (1994) each note that anthropogenic soils—old settlements with soils enriched by manure, ash, garbage, and the like—are sought out as field sites today. Abandoned settlements, or abandoned sectors of occupied villages, could have been farmed during prehistory as well. Denevan also argues that the *extensive* nature of contemporary shifting cultivation systems (i.e., a few years cropping with 20+ years of fallow) is not a good model for the prehistoric situation, but is an artifact of steel implements and reduced population numbers. In other words, shorter fallow is feasible on a sustainable basis in the forested tropics, but is just not commonly practiced today. He thinks that once a clearing was made at great effort using stone tools, it would be kept open and used intensely, with only short fallow. Farmers could also take advantage of natural openings, and seek out secondary growth with softer wood. There is some evidence for the feasibility of this cropping scheme from the Kayapó in eastern Brazil, who use fields for 5–6 years, and then fallow them for 8–11 years (Posey, 1984). Management practices include in-field burning, composting and mulching, and polycropping, as discussed earlier.

House gardens may also have played a role in intensification of cropping. Gardens play a minor role in food production today in the neotropics, largely because houses are frequently moved as field locations shift. If fields were not moved frequently, but grew outward from a core kept in production for long periods, houses could also be more permanent, and gardens larger, more intensively cultivated, and productive. And as Lathrap (1970), among others, has argued, the house garden makes a great experimental plot.

Essentially, then, Denevan is arguing that *extensive* systems of shifting cultivation were too labor-intensive (not less so) to be a common strategy in the tropical lowlands during prehistory, and that most agriculture was semipermanent (short-fallow) or permanent. Polycultural plantings were more protective of soils than monocultural or zonal plantings, and therefore sustainable. As I will discuss further below, this model is testable at least in part through the archaeological record in the Jama Valley: Is there evidence for savanna development (i.e., resulting from unsustainable agricultural practices such as fallow shortening), or is our case an example of 3,600 years of sustainable tropical forest agriculture, in the face of rising populations and increased social and political complexity?

There are environments in the lowland tropics, including the Jama River Valley, with greater potential for intensive agriculture. Let us turn to a brief discussion of floodplain agriculture.

We have few observations of these systems in traditional form, due to depopulation at contact and the attractiveness of these lands to European settlers. And it is difficult to generalize, since floodplains are heterogeneous, and their fertility depends on the geology of the watershed (Moran, 1993). Floodplains do have a predictable form, however, that can help us reconstruct their use (Figure 10.2). In essence floodplains, even small-scale systems like the Jama River, can be conceived of as zonal environments: Next to the active channel, sandbars and mudflats (*playas*) are exposed only at lowest water; in back of these are the high levees, with associated swamp forests on the backslope, inundated only by the highest floods; and further back are low levees, exposed when normal flood levels drop. Landscapes are unstable; as the river moves in, the floodplain land is created in one place, and destroyed in another. Crops can be cultivated annually on *playas* and low levees after floodwater recedes, with some risk of return of high water, but with the benefit of fertile sediments and destruction of vegetation and pests. On higher levees, soils are not annually renewed (although still of recent alluvial origin) and weeds and pests will invade, but floods are less of a hazard. Short-fallow cultivation is most likely in such environments. In Amazonia, villages are typically found on fringing river bluffs, rather than within the floodplain.

Other features of contemporary floodplain environments are river terraces, alluvium laid down in the Pleistocene and Holocene, and, as in the Jama River Valley, now being down-cut by active river channels. River terrace soils are thus relatively young, but are not annually renewed. There are little data, however, on agricultural use of terraces today. In large river systems, old terrace remnants may be far from the active river channel and difficult to access by canoe. In the case of the Jama River Valley, terraces are comprised of sediments laid down during human occupation of the valley (and earlier), and were favored settlement localities in spite of evidence for episodic inundation.

In a floodplain environment, crop mix is matched to the pattern of river fall and rise. Permanent tree crops, such as palms and fruit trees, would be restricted to the highest, rarely flooded ground. Fast-growing, nutrient-demanding crops such as maize, beans, peanuts, tobacco, ajíes, and squashes would be sown on playas and low levees. Higher levees would be good locations for longer-growing-season root crops such as yuca, *achira*, arrowroot, and *llerén*, as well as cotton intercropped with maize.

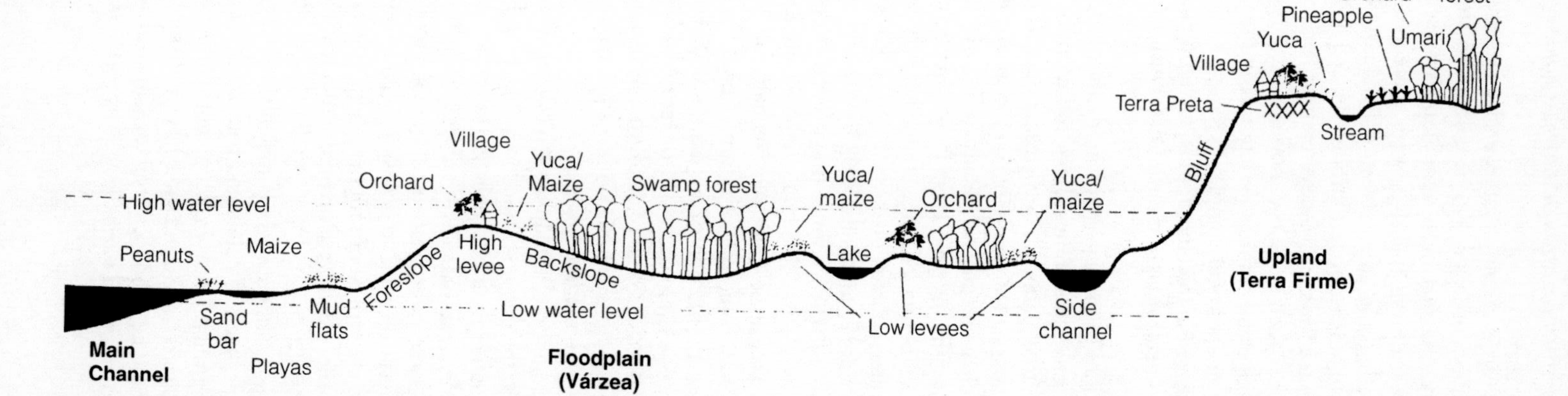

Figure 10.2 River floodplain cross-section. From "A Bluff Model of Settlement in Prehistoric Amazonia" by William Denevan, Annals of the Association of American Geographers *86(4): 654–681. Reproduced with permission from Blackwell Publishers.*

AGRICULTURE IN DRY TROPICAL FORESTS

As discussed in Chapter 3, the "natural" vegetation of the Jama River Valley is classified as very dry and dry tropical forest (coast and Narrows) and dry premontane and humid premontane tropical forest (middle valley zone). Much of the central part of the coastal plain of Ecuador would be characterized by dry forests today—forests in which there are several months of severe, even absolute, drought and less than 1,600 mm of rain per year—if most had not been transformed into plantations and cattle pastures. The moister forest types of the middle Jama Valley result from the often cloud-covered mountain ranges that parallel the coast.

Conversion of tropical dry forests to agriculture and pasture is widespread in the tropics; dry forests are considered among the most endangered ecosystems in the world today (Gentry, 1995) (Figure 10.3). Conversion for plantation agriculture, pasture, subsistence farming, and fuel production has proceeded at such a pace in coastal Ecuador that the remaining dry forest patches are probably too small for viable ecosystem preservation.

Why are dry tropical forests so attractive to humans, especially agriculturalists, both today and in the past? In comparison to humid tropical forests (the rainforests or jungles of popular conception), dry forests are more favorable environments for humans (Maass, 1995). Besides providing people with a break from heat, humidity, and annoying insects and parasites, a lengthy dry season makes burning for field preparation easier and more efficient. Dry forests have a higher ratio of root:above-ground biomass than do humid forests, and root biomass is protected from combustion; therefore, more nutrients are released into the soil by in-situ decay through dry forest burning. The lower above-ground biomass of a dry forest is also easier to cut. A strong dry season reduces overall biological activity in fields, lowering crop pest problems. The shorter growing season reduces the risk of soil degradation resulting from continuous cropping. Further, less energy is expended to protect crops from aggressive competitors (weeds) than in aseasonal environments. Reduced rainfall results in slower loss of nutrients from the soil by leaching, and dry forest soils are less susceptible to compaction.

Dry forests are less diverse in wild plant and animal species than moist forests, however (Gentry, 1995; Valencia et al., 1998). Typically lowland dry forests have 50–70 plant species larger than 2.5 cm in stem diameter per 0.1 hectare, moist semi-evergreen forests 100–150, and wet forests 150–200 species or more. Dry forests of the neotropics share a characteristic suite of families and genera that are in essence a subset of the more diverse floras of humid forests, and that are adapted to seasonal drought. While this difference in diversity might suggest that more kinds of medicinal and other useful wild plants are available for human use in humid forests, in fact regrowing forests are often prime sources of useful plants, as discussed earlier. Further, the defining characteristic of dry tropical forest plants—the ability to survive prolonged drought through dormancy mechanisms—may increase their desirability to humans through the production of starch-rich storage organs (Piperno and Pearsall, 1998).

It is informative to consider the advantages of dry, or seasonal, tropical forests to agriculturalists in light of the issue of how agriculture evolved in the lowland tropics, namely, the proposition that permanent or semipermanent cropping characterized prehistoric systems. Study of contemporary shifting agriculture suggests that

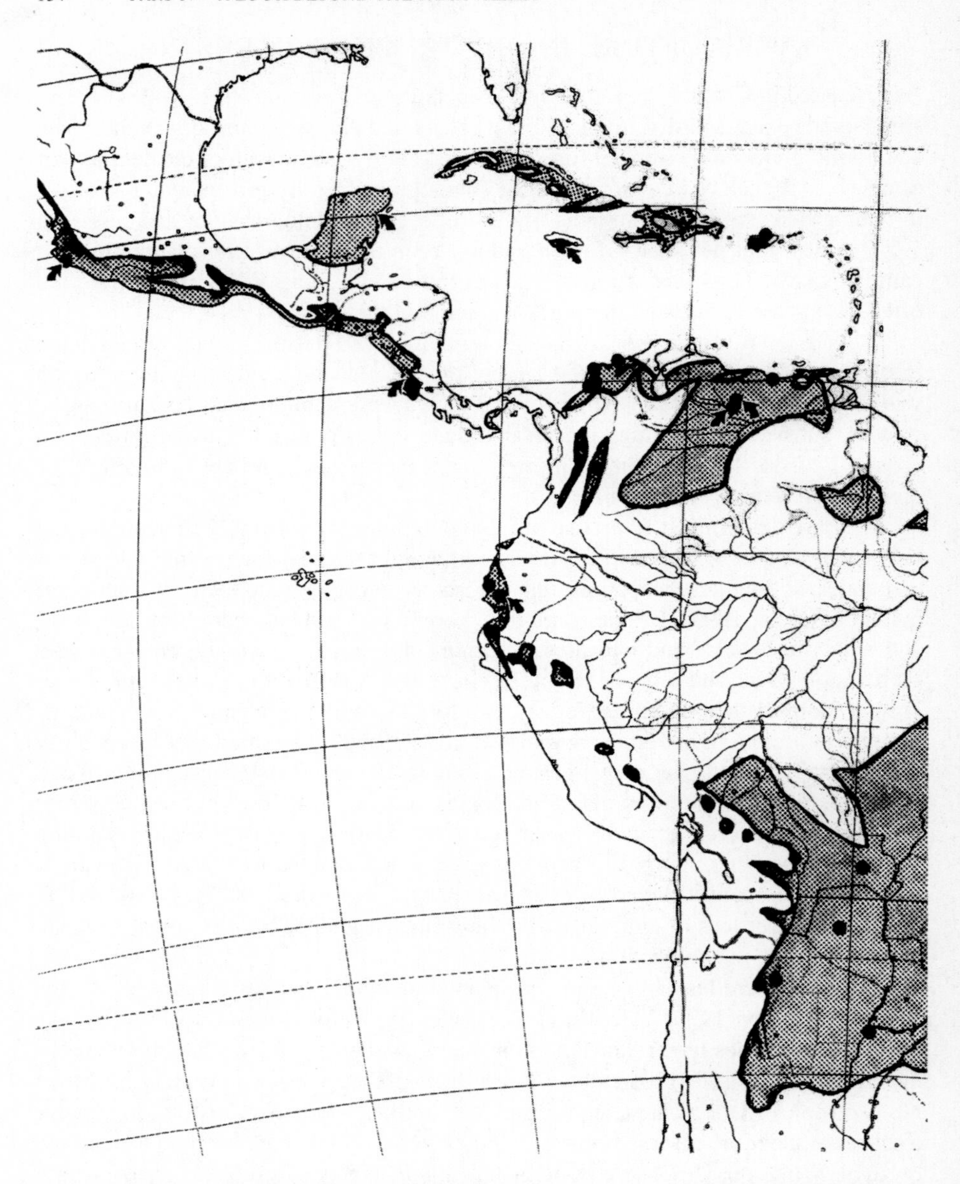

Figure 10.3 Distribution of neotropical dry forests indicated by cross hatching. Adapted from "Diversity and Floristic Composition of Neotropical Dry Forests" by Alwyn H. Gentry, in Seasonally Dry Tropical Forests, *ed. Stephen H. Bullock, et al. Reprinted with the permission of Cambridge University Press.*

invasion of fields by weeds and to a lesser extent, declining fertility are the factors limiting the number of seasons a field can be cropped; therefore, agriculturalists in seasonal forests are at a competitive advantage for extending cropping. Lower weed and pest problems, resulting from dry season die-offs and ease of hot burns, would reduce the labor necessary to weed fields manually and permit longer use. Higher starting fertility rates, resulting from the in-situ release of nutrients from decay of high root biomass, would extend the useful life of fields. While plots in seasonal environments cannot be multicropped in the absence of irrigation, there are advantages to reduced rainfall, namely, slower rates of leaching of soil nutrients, less erosion, and less soil compaction.

We might expect, then, to find evidence for short-fallow shifting cultivation first in dry forested environments, and only later in humid forest settings. Tree felling with stone axes in either environment is time-consuming; in humid forests, increased labor costs for weeding and fertility maintenance might make these habitats considerably less desirable than seasonal forests for agriculturalists.

Would agriculturalists utilizing floodplains in seasonal forests enjoy any advantages over populations using such lands in humid forests? Leaving aside considerations such as availability of game, fish, and other resources, this question is difficult to address given the heterogeneity of river floodplains. Three types of rivers characterize the watersheds of the Amazon, for example (Moran, 1993). "White-water" rivers originate in the Andes and carry sediments of high fertility; "clear-water" rivers head up in the Brazilian and Guiana shields and carry sediments of medium to low fertility from these ancient geological formations; and "blackwater" rivers drain areas of extremely acidic, poor soils. It is also clear that rivers with extensive watersheds like the Amazon present challenges to agriculturalists in terms of timing of planting and loss of crops due to unexpected flooding. That such challenges were met successfully by prehistoric farmers is obvious from the archaeological evidence of sites lining the river bluffs in some parts of the basin. In coastal Ecuador, the large rivers of the interior coastal plain, the Daule, Babahoyo, and Esmeraldas, provide an analogous situation to white-water Amazonian rivers: With Andean headwaters, they run through humid tropical forest and are subject to large-scale seasonal flooding.

Recall that a low-elevation chain of mountains runs parallel to the coast and the Andes in western Ecuador. The coastal cordillera is composed of a core of volcanic and sedimetary rocks of Jurassic and Cretaceous age, with superimposed Tertiary sediments along the flanks, and limited Quaternary deposits along the modern shoreline (uplifted Pleistocene marine terrace) (Zeidler and Kennedy, 1994). As discussed in Chapter 8, a diverse array of soils has developed from this substrate, including several classes of fertile soils. Rivers originating in this chain, including the Jama River, are much smaller in scale than the rivers of the inner coastal plain, both because they drain much more limited terrain and because rainfall is considerably less on an annual basis than in the interior. Some feeder streams dry up seasonally, although the main river channels carry water year-round. Smaller water volume results in smaller-scale floods and limited, sometimes discontinuous, floodplains, as along the Jama channel.

Thus, while annually renewed, fertile alluvial lands may be available to farmers in a small river system, such lands are of limited extent, and even more so in a system whose watershed is in a dry tropical climatic zone. The smaller scale of such river systems, however, reduces the risk factors associated with farming large active floodplains.

To summarize, the following testable hypotheses can be derived from these observations (and speculations) on the nature and evolution of farming systems in the lowland tropics: (1) that prehistoric tropical forest agriculturalists favored cultivation of alluvial lands along small-scale, white-water rivers that ran through dry forest habitats, and that short-fallow swiddening emerged earliest in this setting; (2) that large-scale, white-water rivers in dry forest settings would also be attractive, but would require more sophistication in crop choice and timing; (3) that fertile alluvial lands would attract populations into humid forest settings, but probably after dry forest habitats were heavily used; and (4) that uplands in humid forest settings were the least desirable habitats for tropical forest agriculturalists, the refuge of displaced populations.

EVOLUTION OF TROPICAL FOREST AGRICULTURE

Evidence from the Jama River Valley

The findings of the Jama project, including settlement pattern data and archaeobotanical and vegetational records, support to some extent the model I have outlined for the evolution of tropical forest agriculture. Shifting cultivation does appear fairly early in the Jama sequence, however. Let us review the evidence.

Our single paleoenvironmental record, the Río Grande profile analyzed by Veintimilla (1998, 2000), provides some insights into the sequence of cropping systems in the valley. Recall that during the first phase of occupation in the valley, the early Piquigua phase, there is little reduction in forest cover in comparison to the "preoccupation" zone of the profile (Figure 9.1). Open-area indicators are slightly increased, however; specifically, there are higher levels of canes and other tropical grasses. I infer from these results that there is little forest clearance in the watershed. Since crops are present at sites, it follows that nonforested lands—the alluvial terraces on which Piquigua phase sites were located—were being cropped. This supports Denevan's model.

There is also evidence that some gallery forest fringing the active alluvium was being cleared. Bamboo is a common component of open secondary growth along the Jama River today; in a comparative soil sample taken along the river near Muchique in the middle valley, bamboos and canes made up 34% and 50% of the phytolith assemblage, respectively. Bamboo is virtually absent in the preoccupation zone of the profile, but makes up 7% of the latest early Piquigua sample, suggesting slight opening of gallery forests. This is further support for cropping in riverine habitats.

After the valley is reoccupied following the Tephra I event, there is a significant drop in forest cover in comparison to the earlier levels of the profile. Further, this shift is associated not just with increased bamboo (19%), suggestive of increased gallery forest clearance, but also with the first significant spike in panicoid grass phytoliths (a jump to 14% from earlier levels of less than 6%) and the first occurrence of the forest edge plant *Heliconia*. Since vegetation had likely recovered from the Tephra I ash fall after a hiatus of 425–560 years, these shifts are most likely attributable to human activity, namely, forest-based cultivation. Panicoid phytoliths occur in comparative soil samples from disturbed habitats, being especially common in the drier lower valley. The jump in panicoid phytolith occurrence in late Piquigua-Tabuchila times in the Río Grande profile suggests that land was being opened back from the river edge, in addition to clearance in gallery forests. The first evidence for settlements in nonalluvial areas dates to the Tabuchila phase, and occupation is dense along the Jama River and its major tributaries.

By Muchique 1, following the Tephra II event, the evidence for shifting cultivation is clear. Increased occurrence of bamboos and canes attests to an altered gallery forest; a further jump in panicoid grass occurrence (to 21–25%) and a drop in forest indicators (to 34–36%, the lowest levels for the sequence) document clearance away from the river. The levels dating to Muchique 2 and later, after Tephra III, indicate a slight reversal of these trends, showing an increase in arboreal and wet area indicators, and a drop in panicoid grasses. Appreciable settlement in the uplands is documented for Muchique 2 and after.

Can we draw any inferences about length of cropping and fallow cycles from these data? Comparative soil samples from the middle valley indicate that "mature" forest (in this case long-growth secondary forest with nearly closed tree canopy) has a background level of 7–8% panicoid grass phytoliths (Veintimilla, 1998). Preoccupation levels of the Río Grande profile have no more than 6% panicoids. These data suggest, then, that panicoid phytolith levels higher than 8% indicate a more open forest. A comparative sample from the Cerro Sancho area in the middle valley, for example, has 24% panicoid grasses and 43% total arboreal phytoliths, with canopy tree cover estimated at 50%. By Muchique 1 times and continuing into Muchique 2, phytolith assemblages of prehistoric samples are very similar to the pattern of the Cerro Sancho comparative soil. From this I infer that by Jama-Coaque times, the watershed of the middle valley was characterized by large expanses of open, secondary forest, with grasses being a significant component of the understory. The continued presence of a forest component suggests a short-fallow pattern (8–10 years).

Was the agroecology of the middle valley deteriorating into savannah, or had it stabilized as short fallow? Without having paleoenvironmental samples from the later phases of occupation of the valley, it is difficult to say. One indication of stability is the fact that panicoid grass levels actually fall slightly between the early and later Muchique 2 samples, rather than increase. Quantitative data on the occurrence of particulate charcoal in the samples would be helpful in assessing the extent to which environmental disturbance was increasing late in the sequence. We do know from observations of 20th-century settlers to the valley, and travelers such as Saville (1907, 1910), that forest had regrown during the centuries the valley was largely abandoned.

Macroremain and phytolith data from archaeological sites tested during the project, especially the well-preserved record from the Big Pit at the Pechichal site, provide some evidence that Jama farmers practiced polycropping. The diverse array of foods grown or gathered from tended wild stands suggests that mixed plantings were the rule. As discussed in Chapter 9, however, it would make sense for limited alluvial lands to be used for producing maize, or crops favored by elites, or both.

In summary, the Jama River case supports the model that agriculturalists moving into a new habitat, and dependent on stone-tool technology, will favor use of annually renewed lands requiring minimal clearing. Early occupation of the valley is clearly oriented toward the largest pockets of alluvium along the Jama River and its major tributaries (Pearsall and Zeidler, 1994), and there is no evidence for shifting cultivation. Further, there is some indication that once agricultural plots were opened in upland forest areas, these were maintained under a short-fallow regime. In other words, we do not see a gradual drop in forest cover that might be expected if plots were allowed to regrow for 20+ years. Rather, arboreal indicators drop fairly abruptly to levels suggesting open canopy forests. This drop occurs before there is evidence for extensive habitation in the uplands: Farmers were not "forced" to shorten fallow in response to population pressure, but chose to keep fields in cultivation because this was labor-efficient. The mix of crops present suggests polycropping was practiced.

The Jama data do not allow us to address in a satisfactory way the question, Was this agricultural system ultimately sustainable? To test whether the cropping practices could have been sustained indefinitely without ongoing deterioration of environment (e.g., soil erosion, savanna invasion, salinization), and at what population level, requires fine-grained, well-dated paleoenvironmental sequences from throughout the valley, closely tied to the archaeological sequence. Future research may allow us to do such sampling. For now, the Río Grande profile does document a *drop* in panicoid grass levels between the early and later Muchique 2 samples, rather than the increases that savanna invasion would bring. This is only one datum, however, and it is from relatively early in the sequence.

Zeidler's analyses to date of site testing and pedestrian survey data indicate that political power shifted from San Isidro, in the middle valley, to the lower valley late in prehistory (Pearsall and Zeidler, 1994). If soils did deteriorate in long-cropped upland areas to the southwest of San Isidro, this could have played a role in the shift of power. As I demonstrated in Chapter 8, the lower valley is also very productive for maize agriculture, and is close to potential sources of fertilizer: guano deposits and fish.

It is beyond the scope of this book for me to review other paleoenvironmental sequences from the neotropics relevant for investigating the nature of agricultural evolution in the lowland tropics. In a number of sequences from Central and South America with which I am familiar (Pearsall, 1995b; Pearsall and Jones, 2001), it is possible to detect the advent of shifting cultivation—the increase in particulate charcoal in the record is often dramatic—and to track the subsequent course of deforestation. But to determine whether crops were being grown *before* shifting cultivation was practiced (i.e., on annually renewed lands) is more difficult, since this requires finding evidence for the crops themselves in association with "natural" vegetation. Identifying what "natural" is for the early Holocene, finding the crops themselves (which may not yet exist in modern form), and providing convincing dates are just some of the challenges. It sounds like fun, though, doesn't it?

11/Ethnobotany in Archaeology

WHAT KIND OF ARCHAEOLOGY IS THIS, ANYWAY?

My goal in writing *Plants and People in Ancient Ecuador: The Ethnobotany of the Jama River Valley* was to explore the interrelationships between the prehistoric populations of the valley and the dry tropical forest habitat in which they lived for some 3,600 years. I did this by studying the direct material correlates of those interrelationships—charred macroremains produced by cooking accidents, burning of wood as fuel, or refuse disposal—and phytoliths deposited naturally by the flooding of the Jama River or through the decay of plants brought to and used at sites. To better understand how Jama-Coaque populations may have produced food in the valley, I investigated the productivity of two important crops, maize and yuca, under traditional agricultural practices.

I summarize in the next section the major findings of the ethnobotanical studies of the Jama project, and suggest some directions for future research. First I would like to return to some of the themes of Chapter 1, and consider in general terms how this kind of archaeology contributes to our understanding of past cultures.

I started Chapter 1 with quotations from Schultes and von Reis, and Balick and Cox that speak in different ways about the importance of ethnobotany. Knowledge of useful plants has not only played an essential role in the success of *Homo sapiens,* but has also enriched human cultures, both traditional and modern. The study of archaeological plant remains gives us a window, albeit a clouded one, into the incredible diversity of knowledge and practice of plant use by cultures that no longer exist. Paleoethnobotany thus contributes to our understanding of how these cultures solved some of the basic problems of life—finding or producing food, creating shelter, maintaining health—and in doing so how they changed the land and the plants and animals that lived on it.

I also describe in Chapter 1 how archaeological projects today commonly incorporate more than one approach for investigating past plant-people interrelationships. In other words, as in the Jama project, macroremains, pollen, and phytoliths—as well as faunal remains and other biological indicators—are often targeted for systematic recovery and analysis. All this costs money and time in the field and laboratory. How does having multiple data sources contribute to understanding past cultures?

There are numerous examples from the Jama project that support the position that multiple lines of evidence are desirable for understanding past plant-people interrelationships more completely. Take the question of use of root and tuber resources. If I had relied only on charred macroremains to document plant use, only two kinds of root or tuber remains would have been identified: yuca (from the "Big Pit") and a monocot rhizome. Other macroremains resembling root or tuber tissues were found, but could not be identified further because of their small sizes and lack of diagnostic features. Through phytolith analysis, however, I was able to document some of the other root or tuber resources that were used: arrowroot, *llerén,* and *achira.* Neither phytoliths nor macroremains alone would have given as complete a picture.

The environmental studies carried out by Veintimilla (1998, 2000) and Stahl (2000) provide an example of how multiple data sources permit a robust reconstruction of past human-landscape interactions. Recall how Veintimilla's phytolith study (the Río Grande profile) provided a look at regional patterns of vegetation change in the valley—the onset of forest clearance and the maintenance of open spaces. Stahl's study of the fauna of the Big Pit at the Pechichal site documented the presence of animals favoring forest edges and open habitats as well as animals expected to thrive in forest fragments, the same types of habitats indicated in the Río Grande profile for Muchique 2 times. Finally, the assemblages of macroremains and phytoliths from the Big Pit make perfect sense in the context of this vegetation reconstruction. The concordance of these independent lines of evidence gives us increased confidence in the vegetation reconstruction, and the implications that it has for evolution of agriculture in the valley.

In the Chapter 1 section entitled *A Few Theoretical Issues,* I describe the paleoethnobotanical research process as one of developing models of past human-plant interrelationships. You, the reader, may or may not find a model convincing, but a good model generates hypotheses whose predictions can be tested in future independent research. Paleoethnobotanical data can also be gathered to test existing models. The Jama ethnobotanical studies illustrate each process, model development and testing. How can this be?

Jim Zeidler and I began the Jama project with the goal of investigating the emergence and evolution of complex societies in northern Manabí province, Ecuador, and the agricultural system that sustained them. In our grant proposals, Jim developed alternative explanations for the emergence of social and political complexity, and described how we would use the data from the archaeological survey and site testing program to evaluate these explanations. I derived hypotheses from one model of agricultural evolution, the coevolutionary model of David Rindos (1984), which I hoped we could test with the Jama paleoethnobotanical data. I have not discussed these models in detail (see Chapter 2 for a brief description of these project goals), nor have I examined how they hold up in light of the Jama data: We are not yet at the point of integrating all the data sets to do this evaluation. What I have done is describe and interpret the paleoethnobotanical record from the valley, in particular the data from the Muchique 2 phase. The *description,* the lists of plants and the quantities of remains that were left, provides the raw data we will eventually use to test Rindos's ideas on the evolution of agriculture and to examine economic relationships within the valley; my *interpretation,* that the Jama case is an example of sustainable dry tropical forest agriculture, creates a model of past plant-human interrelationships that future research may uphold or reject.

LESSONS AND FUTURE DIRECTIONS OF THE JAMA PROJECT

What did we learn about this "gray area" of the archaeological map of Ecuador through the analysis of phytoliths and macroremains from 14 sites in the Jama River Valley? The major findings are:

1. A diverse array of economically useful plants was used in the valley. Many of these are cultivated plants (those planted or otherwise tended by people), including domesticated species (genetically changed through human selection from wild related plants). These include root or tuber-producing plants, such as *achira,* arrowroot, *llerén,* and yuca; a legume, the common bean; the most important New World grain, maize; gourd and possibly squash; and cotton.
2. Maize and bean remains were common enough at the Pechichal site, and to a lesser extent at El Tape, to allow some characterization of the varieties present. The maize is a small-kerneled, somewhat variable 16-row variety that was grown in both the lower and middle valley in Muchique 2 times. It produces smaller kernels (and likely smaller ears) than either of the two surviving traditional coastal varieties. Phytoliths produced by cob remains exhibit some primitive characteristics. Beans are also small, but in the size range of other archaeological common beans recovered from South American sites, especially if shrinkage during charring is taken into account.
3. A number of kinds of tree fruits were used by inhabitants of the Jama Valley. Some of these may have been tended or encouraged, or perhaps increased naturally in areas of human disturbance. Among these are soursop, *cadi* palm, *coroso* palm (as well as unidentified palms), tree gourd, *achiote, pechiche,* tree legumes, *guava, sapote,* and 10 unidentified arboreal morpho-types.
4. Seeds of plants favoring open or disturbed habitats were present in samples, but rarely in abundance. Phytoliths from open-area plants were a common component in all the sequences, however. Interpreting these indicators is difficult. Do the charred remains and phytoliths document human use of sun-loving plants (e.g., as greens, medicines, or spices), or are they the background signal from disturbed habitats around sites or secondary forest in the watershed of the river?
5. For the Muchique 2 phase, macroremains and phytoliths reveal a subsistence base that included domesticated, tended, and wild plant species; root/tuber crops and seed crops; and annual and perennial resources. The well-preserved Big Pit assemblage documents the richness of utilized plants. There is no evidence for a starchy or oily "small seed complex," however. While greens and fruits of plants growing in disturbed and open habitats were available seasonally and likely used, there are no concentrations of charred seeds suggesting that such plants were a focus of collecting or storage. There is a hint of specialization in crop production at Pechichal and El Tape.
6. My study of maize and yuca productivity in the valley today revealed that there are three broad classes of "maize lands," from highest to lowest yielding:

river alluvium, mollisols (two levels of productivity), and inceptisols. Agricultural productivity is not uniform around the valley today, and would not have been so in the past. Better maize-producing soils dominate the higher-elevation areas to the south of San Isidro in the middle valley; the town (and site) itself sits in one of the largest pockets of rich alluvial land. These favorable circumstances may have contributed to the early emergence of San Isidro as the center of political power in the valley. Yuca, however, is the unknown variable: It may have played the role of "equalizer" in crop productivity around the valley, since it produces equally well on all aerable soils.

7. The Río Grande phytolith profile documents the emergence of an agroecology, a landscape managed by people for crop production, over the course of some 4,000 years. By Muchique 2 times, expanses of land were maintained in increasingly open conditions. Animals trapped in the Big Pit are those that thrive in secondary forest, forest fragments and edges, and agricultural fields.
8. Three episodes of volcanic ash fall are documented in profiles around the valley. The final episode, Tephra III, separates the Muchique 1 and Muchique 2 phases, but did not lead to a hiatus in occupation of the valley. Existing upland fields would have quickly rebounded as ash eroded into the valley bottoms, and new fields were easily created in areas of high tree mortality. While perennial crops would have been destroyed, maize was well adapted to these conditions. Populations occupying valley lands, rendered unusable for agriculture, did not leave for any length of time, and so must have had access to upland plots or been provisioned by other members of the society (perhaps through family ties or a network of formal redistribution of resources).
9. The Jama project results suggest that agriculturalists first utilized alluvial lands in the valley, and only later cleared fields in the forested uplands. The pattern that emerges for upland cultivation seems to be one of maintenance of open landscapes, rather than long-fallow shifting cultivation. The natural advantages of dry tropical forests for agriculturalists, and the difficulty of clearing forest with stone tools, may explain this pattern, which runs counter to models based on contemporary tropical forest agriculturalists.

What of the future? Ahead lies the exciting prospect of integrating the results of the settlement survey, site testing program, artifact analyses, and floral and faunal studies to test the "big issues" that drove the Jama Valley Archaeological-Ethnobotanical Project, and to describe the 3,600-year history of human occupation of this small coastal valley. Ahead also lies the possibility of continued fieldwork in the region: exploration of adjoining valleys that were part of the Jama-Coaque polity; expanded horizontal excavations at selected sites to investigate intrasite patterning and to increase the data base of biological samples; environmental coring in the lower and middle valley to refine the sequence of forest clearance, better documentation of the effects of tephra fall, and the testing of our ideas on the evolution of Jama-Coaque agriculture. Many of these goals will be advanced using the techniques and approaches of paleoethnobotany, the study of plant-people interrelationships through the archaeological record.

Appendix

Macroremains from the Pechichal Site, Excluding Unknowns

Cateo	1	1	1								
Feature				5	5	5	5	5	5	5	5
Deposit/Element				A	B	C	D	E	F	G	H
Level	20-40	60-80	80-100								
Liters	30	30	30	4.5	44	14	28	54.5	21	28.5	80
Catalog #	021	023	024	025	026	027	028	029	034	035	036
Flotation #	89-5	89-6	89-7	89-22	89-11	89-23	89-10	89-9, 89-21	89-17	89-25	89-12, 89-13
Wood, Ct.	4	83	24	18	167	89	198	371	63	215	317
Wood, Wt.	0.1141	1.3221	0.2004	0.1789	1.2099	3.9142	2.2991	3.7823	0.5319	1.4518	2.9637
Small diam. stems, Ct.											
Small diam. stems, Wt.											
Cultivated											
Maize kernel frag., Ct.	2	17	7	6	19	21		79	9	29	55
Maize kernel frag., Wt.	0.0614	0.0682	0.0402	0.0386	0.0678	0.2324		0.4408	0.0358	0.1628	0.3534
Maize cupule frag., Ct.	5	66	43	13	127	10	3	37	11	237	492
Maize cupule frag., Wt.	0.0057	0.0318	0.0922	0.0988	0.2927	0.0482	0.0052	0.0973	0.0462	0.2663	1.0405
cf. Manihot esculenta, Ct.											
cf. Manihot esculenta, Wt.											
Monocot rhizome, Ct.											
Monocot rhizome, Wt.											
Root/tuber frag., Ct.					5			2	2	59	119
Root/tuber frag., Wt.					0.0237			0.0435	0.1422	0.4104	1.0487
Gossypium, Ct.		3	1		1			4	2		4
Gossypium, Wt.		0.0216	0.0028		0.0098			0.0347	0.0216		0.0267
Phaseolus frag., Ct.			2	2	2	4		16		8	10
Phaseolus frag., Wt.			0.0177	0.0147	0.024	0.0942		0.2088		0.0685	0.2014
Lagenaria rind, Ct.										1	5
Lagenaria rind, Wt.										0.0007	0.0035

MACROREMAINS FROM THE PECHICHAL SITE, EXCLUDING UNKNOWNS (*continued*)

Cateo							
Feature	5	5	5	5	5	5	5
Deposit/Element	I	J	K	L	M	N	N
Level							
Liters	6	27.5	21	86	1.5	12.5	34
Catalog #	037	038	039	040	041	042	045
Flotation #	89-15	89-16, 89-19	89-18	89-14, 89-32	89-20	89-26	89-31
Wood, Ct.	52	141	9	174	11	56	199
Wood, Wt.	0.414	1.5331	0.0221	2.07	0.0886	0.6873	5.1575
Small diam. stems, Ct.				1			
Small diam. stems, Wt.				0.0067			
Cultivated							
Maize kernel frag., Ct.	659	79	194	91	22	105	30
Maize kernel frag., Wt.	8.0724	0.7474	2.0136	0.4321	0.129	0.2338	0.3409
Maize cupule frag., Ct.	25	10	7	78		131	88
Maize cupule frag., Wt.	0.0322	0.031	0.0154	0.2781		0.2503	0.2196
cf. Manihot esculenta, Ct.							
cf. Manihot esculenta, Wt.							
Monocot rhizome, Ct.						1	1
Monocot rhizome, Wt.						0.007	0.0116
Root/tuber frag., Ct.	61	6	4	7		157	144
Root/tuber frag., Wt.	0.7686	0.0377	0.0916	0.0901		1.4107	2.1686
Gossypium, Ct.	6	3		29	4	5	117
Gossypium, Wt.	0.069	0.0238		0.1501	0.039	0.046	1.4408
Phaseolus frag., Ct.	50	10	8	4	7	52	106
Phaseolus frag., Wt.	1.2993	0.2792	0.1716	0.0962	0.1158	0.8174	1.8542
Lagenaria rind, Ct.						1	
Lagenaria rind, Wt.						0.0016	

Cateo								Totals
Feature	5	5	5	5	5	5	5	
Deposit/Element	L (Q?)	P	Q	R	R	S	T	
Level								
Liters	18	54	108	27	37.5	27.5	9	834
Catalog #		044	045		046		048	
Flotation #	89-29	89-28, 89-30	89-27, 89-36, 89-37	89-34	89-35	89-33	89-24	
Wood, Ct.	65	51	224	40	146	152	33	2902
Wood, Wt.	0.5962	1.6909	3.0734	0.3062	4.2586	0.6981	0.2002	38.7646
Small diam. stems, Ct.								1
Small diam. stems, Wt.								0.0067
Cultivated								
Maize kernel frag., Ct.	12	72	153	37	46	16	2	1762
Maize kernel frag., Wt.	0.0713	0.4023	1.5816	0.2975	0.2537	0.0724	0.0096	16.159
Maize cupule frag., Ct.	13	246	574	10	104	8	2	2340
Maize cupule frag., Wt.	0.1251	0.3107	2.062	0.0418	0.4233	0.0102	0.0135	5.8381
cf. Manihot esculenta, Ct.	249		569					818
cf. Manihot esculenta, Wt.	3.1183		5.8339					8.9522
Monocot rhizome, Ct.								2
Monocot rhizome, Wt.								0.0186
Root/tuber frag., Ct.		8			15		1	590
Root/tuber frag., Wt.		0.2743			0.1306		0.0252	6.6659
Gossypium, Ct.	103	5	68	5		3		363
Gossypium, Wt.	0.9845	0.0554	0.7065	0.0258		0.0183		3.6764
Phaseolus frag., Ct.	24	48	284	2	9		3	651
Phaseolus frag., Wt.	0.404	0.9847	6.4358	0.0349	0.1728		0.0093	13.3045
Lagenaria rind, Ct.						1		8
Lagenaria rind, Wt.						0.0014		0.0072

MACROREMAINS FROM THE PECHICHAL SITE, EXCLUDING UNKNOWNS (*continued*)

Deposit/Element				A	B	C	D	E	F	G	H
Level	20-40	60-80	80-100								
Arboreal											
Aiphanes, Ct.		1									
Aiphanes, Wt.		0.0236									
Phytelephas, Ct.											
Phytelephas, Wt.											
Arecaceae frag., Ct.		2	3		3	6		16	1	10	21
Arecaceae frag., Wt.		0.0982	0.4375		0.0174	0.0832		0.0968	0.0774	0.0345	0.0199
Bixa orellana, Ct.						1		5	1		28
Mimosaceae (tree legume), Ct.											
Psidium, Ct.											
Sapotaceae seed frag., Ct.								20		3	24
Sapotaceae seed frag., Wt.								0.012		0.0053	0.0161
cf. Sideroxylon, Ct.			1	3	5	10		16		3	18
Thick rind, Ct.					64	12		62	21	6	1
Thick rind, Wt.					0.5171	0.1009		0.4694	0.1428	0.0485	0.0056
Thick flat rind, Ct.											
Thick flat rind, Wt.											
Thick rind, internal div, Ct.											
Thick rind, internal div, Wt.											
Thin rind, Ct.											
Thin rind, Wt.											
Thin rind w/storied str., Ct.								16			6
Thin rind w/storied str., Wt.											
Curved rind, Ct.											
Curved rind, Wt.											

Deposit/Element	I	J	K	L	M	N	N
Level							
Arboreal							
Aiphanes, Ct.							
Aiphanes, Wt.							
Phytelephas, Ct.		11		5			1
Phytelephas, Wt.		1.9354		0.0443			0.01
Arecaceae frag., Ct.	4	25	6	10	3	10	25
Arecaceae frag., Wt.	0.0891	0.0773	0.014	0.0926	0.0071	0.0556	0.2526
Bixa orellana, Ct.	18	30	7	3	6		
Mimosaceae (tree legume), Ct.	1						
Psidium, Ct.							
Sapotaceae seed frag., Ct.	1		2				1
Sapotaceae seed frag., Wt.	0.0006		0.0704				0.0016
cf. Sideroxylon, Ct.	33	28	3	25	2	51	29
Thick rind, Ct.			8	15			
Thick rind, Wt.			0.0676	0.1052			
Thick flat rind, Ct.							
Thick flat rind, Wt.							
Thick rind, internal div, Ct.							3
Thick rind, internal div, Wt.							0.028
Thin rind, Ct.	9			15			
Thin rind, Wt.	0.0073			0.0179			
Thin rind w/storied str., Ct.				2			
Thin rind w/storied str., Wt.				0.0138			
Curved rind, Ct.							
Curved rind, Wt.							

MACROREMAINS FROM THE PECHICHAL SITE, EXCLUDING UNKNOWNS (*continued*)

Deposit/Element	L (Q?)	P	Q	R	R	S	T	Total
Level								
Arboreal								
Aiphanes, Ct.								1
Aiphanes, Wt.								0.0236
Phytelephas, Ct.			21					38
Phytelephas, Wt.			0.2791					2.2688
Arecaceae frag., Ct.	5	12	28	8	8	22		228
Arecaceae frag., Wt.	0.1022	0.0571	0.1891	0.0623	0.0305	0.4268		2.3212
Bixa orellana, Ct.			11			2		112
Mimosaceae (tree legume), Ct.		2						3
Psidium, Ct.		1	2			1	1	5
Sapotaceae seed frag., Ct.		2						53
Sapotaceae seed frag., Wt.		0.0017						0.1077
cf. Sideroxylon, Ct.		71	17	15	18	4	1	353
Thick rind, Ct.		7			1			197
Thick rind, Wt.		0.0911			0.0652			1.6134
Thick flat rind, Ct.			73		14			87
Thick flat rind, Wt.			0.4329		0.0787			0.5116
Thick rind, internal div, Ct.			40		1			44
Thick rind, internal div, Wt.			1.6476		0.0322			1.7078
Thin rind, Ct.								24
Thin rind, Wt.								0.0252
Thin rind w/storied str., Ct.								24
Thin rind w/storied str., Wt.								0.0138
Curved rind, Ct.		3						3
Curved rind, Wt.		0.0229						0.0229

Deposit/Element				A	B	C	D	E	F	G	H
Level	20-40	60-80	80-100								
Dense cotyledon frag., Ct.	8	3	7			2		71		9	63
Dense cotyledon frag., Wt.	0.1212	0.0259	0.19			0.0149		0.3062		0.0355	0.3061
Small dense spherical fruit, Ct.			1		2						
Small dense spherical fruit, Wt.			0.0685		0.0435						
Dimpled fruit peel, Ct.											
Dimpled fruit peel, Wt.											
Unk 82	1										
Unk 83											
Unk 87, Ct.											
Unk 87, Wt.											
Unk 104, Ct.											
Unk 104, Wt.											
Porous endosperm frag., Ct.		1	6					3			7
Porous endosperm frag., Wt.		0.0068	0.0648					0.0196			0.0146
Small Seeds											
Amaranthus								4			11
Phytolaccaceae											
cf. Iresine								11			11
Asteraceae											
Fabaceae (744)											
Fabaceae, beaked											
Fabaceae, elongated											
Fabaceae, small								1			
Fabaceae, medium					1						

MACROREMAINS FROM THE PECHICHAL SITE, EXCLUDING UNKNOWNS (*continued*)

Deposit/Element	I	J	K	L	M	N	N
Level							
Dense cotyledon frag., Ct.	17	10	11	9	5	9	45
Dense cotyledon frag., Wt.	0.115	0.0821	0.0853	0.0282	0.0192	0.0361	0.3113
Small dense spherical fruit, Ct.							2
Small dense spherical fruit, Wt.							0.0404
Dimpled fruit peel, Ct.							
Dimpled fruit peel, Wt.							
Unk 82							
Unk 83				1			
Unk 87, Ct.							2
Unk 87, Wt.							0.0662
Unk 104, Ct.							
Unk 104, Wt.							
Porous endosperm frag., Ct.	9					4	67
Porous endosperm frag., Wt.	0.041					0.3114	0.2532
Small Seeds							
Amaranthus	1				2	2	
Phytolaccaceae							
cf. Iresine							
Asteraceae							
Fabaceae (744)							
Fabaceae, beaked						2	
Fabaceae, elongated						1	
Fabaceae, small						4	
Fabaceae, medium						1	2

Deposit/Element	L (Q?)	P	Q	R	R	S	T	Total
Level								
Dense cotyledon frag., Ct.		22	153	9	16	10	2	481
Dense cotyledon frag., Wt.		0.1022	1.3627	0.0578	0.2132	0.0913	0.0068	3.511
Small dense spherical fruit, Ct.			14					19
Small dense spherical fruit, Wt.			0.1143					0.2667
Dimpled fruit peel, Ct.		1						1
Dimpled fruit peel, Wt.		0.3425	0.4264					0.7689
Unk 82								1
Unk 83					1			2
Unk 87, Ct.								2
Unk 87, Wt.								0.0662
Unk 104, Ct.			41					41
Unk 104, Wt.			0.7751					0.7751
Porous endosperm frag., Ct.			82				2	181
Porous endosperm frag., Wt.			0.3983				0.0054	1.1151
Small Seeds								
Amaranthus		4	3	13	2	1		43
Phytolaccaceae				160	12	3		175
cf. Iresine								22
Asteraceae					1			1
Fabaceae (744)			1					1
Fabaceae, beaked								2
Fabaceae, elongated								1
Fabaceae, small		3	1					9
Fabaceae, medium								4

MACROREMAINS FROM THE PECHICHAL SITE, EXCLUDING UNKNOWNS (*continued*)

Deposit/Element				A	B	C	D	E	F	G	H
Level	20-40	60-80	80-100								
Fabaceae, large					1						
cf. Linum	5		2		2	1		5			6
cf. Abutilon											
cf. Herissantia	1				1						1
Sida								2			3
Passiflora			1								
Poaceae, small-medium elongate	1	4	2					3			18
Portulaca					1						
cf. Rubus											
Solanaceae	2	3	2	1	1			12		8	6
Teucrium								1			3
Trianthema portulacastrum		11	20	1	37	9		47	4	14	23

Deposit/Element	I	J	K	L	M	N	N
Level							
Fabaceae, large							1
cf. Linum	5	4	3	2		4	4
cf. Abutilon						1	
cf. Herissantia				1	1	2	3
Sida	5	1				1	
Passiflora						1	
Poaceae, small-medium elongate	6			4	1	3	
Portulaca				1			
cf. Rubus					1		
Solanaceae	1	1	26	1		3	4
Teucrium							
Trianthema portulacastrum	1	3		6		1	

MACROREMAINS FROM THE PECHICHAL SITE, EXCLUDING UNKNOWNS (*continued*)

Deposit/Element	L (Q?)	P	Q	R	R	S	T	Total
Level								
Fabaceae, large								2
cf. Linum				13	2		1	59
cf. Abutilon					3			4
cf. Herissantia	1	1	7	30	23	22		94
Sida								12
Passiflora		1	4	9	3			19
Poaceae, small-medium elongate		9		4				55
Portulaca		3		1				6
cf. Rubus							2	3
Solanaceae		6		8			24	109
Teucrium								4
Trianthema portulacastrum		1	1	7		2		188

Bibliography

Acosta-Solis, Misael. 1961. *Los Bosques del Ecuador y sus Productos*. Quito, Editorial Ecuador.

Alexiades, Miguel N. 1996. Standard techniques for collecting and preparing herbarium specimens. Pp. 99–126 in *Selected Guidelines for Ethnobotanical Research: A Field Manual*. Miguel N. Alexiades, editor. The New York Botanical Garden, New York.

Anderson, Anthony B., and Darrell A. Posey. 1989. Management of a tropical scrub savanna by the Gorotire Kayapó of Brazil. *Advances in Economic Botany* 7: 159–173. Publication 32, The New York Botanical Garden, Institute of Economic Botany.

Balée, William. 1994. *Footprints of the Forest. Ka'apor Ethnobotany—The Historical Ecology of Plant Utilization by an Amazonian People*. Columbia University Press, New York.

Balick, Michael J., and Paul Alan Cox. 1996. *Plants, People, and Culture: The Science of Ethnobotany*. Scientific American Library, New York.

Boom, Brian M. 1989. Use of plant resources by the Chácobo. *Advances in Economic Botany* 7: 78–96. Publication 32, The New York Botanical Garden, Institute of Economic Botany.

Bruhns, Karen O. 1994. *Ancient South America*. Cambridge University Press, Cambridge, UK.

Buol, S. W., F. D. Hole, R. J. McCracken, and R. J. Southard. 1997. *Soil Genesis and Classification*. Fourth Edition. Iowa State University Press, Ames.

Castner, James L., Stephen L. Timme, and James A. Duke. 1998. *A Field Guide to Medicinal and Useful Plants of the Upper Amazon*. Feline Press, Gainesville, FL.

Clark, J. G. D. 1954. *Excavations at Star Carr: An Early Mesolithic Site at Seamer, Near Scarborough, Yorkshire*. Cambridge University Press, London.

Cotton, C. M. 1996. *Ethnobotany: Principles and Applications*. John Wiley and Sons, Chichester, UK.

Denevan, William M. 1996. A bluff model of settlement in perhistoric Amazonia. *Annals of the Association of American Geographers* 86(4): 654–681.

Denevan, William M. 2001. *Cultivated Landscapes of Native Amazonia and the Andes*. Oxford University Press, Oxford, UK.

Donahue, Jack, and William Harbert. 1994. Fluvial history of the Jama River drainage basin. Pp. 43–57 in *Regional Archaeology in Northern Manabí, Ecuador, Vol. 1. Environment, Cultural Chronology, and Prehistoric Subsistence in the Jama River Valley*. James A. Zeidler and Deborah M. Pearsall, editors. University of Pittsburgh Memoirs in Latin American Archaeology, No. 8. Pittsburgh, PA.

Duke, James A. 1992. *Handbook of Edible Weeds*. CRC Press, Boca Raton, FL.

Eggler, W. A. 1948. Plant communities in vicinity of the volcano Paricutín, Mexico, after 2 1/2 years of eruption. *Ecology* 29:415–436.

Eggler, W. A. 1963. Plant life of Paricutín volcano, Mexico, 8 years after activity ceased. *American Midland Naturalist* 69:38–68.

Estrada, Emilio. 1957. *Prehistória de Manabí*. Publicación del Museo Víctor Emilio Estrada, No. 4. Guayaquil.

Ferreyra, Ramon. 1970. *Flora Invasora de los Cultivos de Pucallpa y Tingo María*. Gráfica Morsom, Lima, Peru.

Gentry, Alwyn H. 1993. *A Field Guide to the Families and Genera of Woody Plants of Northwest South America (Colombia, Ecuador, Peru) with supplementary information on herbaceous taxa*. Conservation International, Washington, DC.

Gentry, Alwyn H. 1995. Diversity and floristic composition of neotropical dry forests. Pp. 146–194 in *Seasonally Dry Tropical Forests*. Stephen H. Bullock, Harold A. Mooney, and Ernesto Medina, editors. Cambridge University Press, Cambridge, UK.

Goldammer, Johann Georg. 1992. Tropical forests in transition. Ecology of natural and anthropogenic disturbance processes—An introduction. Pp. 1–16 in *Tropical Forests in Transition. Ecology of Natural and Anthropogenic Disturbance Processes*. J. G. Goldammer, editor. Birkhäuser Verlag, Basel.

Harling, Gunnar, and Benkt Sparre (series editors). 1973–1998. *Flora of Ecuador*. Vols. 1–60 (continuing). Department of Systematic Botany, University of Göteborg and Section for Botany, Riksmuseum, Stockholm.

Hastorf, Christine, and Virginia Popper (editors). 1988. *Current Paleoethnobotany. Analytical Methods and Cultural Interpretations of Archaeological Plant Remains*. The University of Chicago Press, Chicago.

Hole, Frank, Kent V. Flannery, and James A. Neely (editors). 1969. *Prehistory and Human Ecology of the Deh Luran Plain*. Memoirs of the Museum of Anthropology, No. 1. University of Michigan, Ann Arbor.

Isaacson, John S. 1994. Volcanic sediments in archaeological contexts from western Ecuador. Pp. 131–140 in *Regional Archaeology in Northern Manabí, Ecuador, Vol. 1. Environment, Cultural Chronology, and Prehistoric Subsistence in the Jama River Valley*. James A. Zeidler and Deborah M. Pearsall, editors. University of Pittsburgh Memoirs in Latin American Archaeology, No. 8. Pittsburgh, PA.

Isaacson, John S., and James A. Zeidler. 1999. Accidental history: Volcanic activity and the end of the Formative in northwestern Ecuador. Pp. 41–72 in *Actividad Volcánica y Pueblos Precolombianos en el Ecuador*. Patricia Mothes, editor. Abyayala, Quito, Ecuador.

Jørgensen, Peter M., and Susana León-Yánez (editors). 1999. *Catalogue of the Vascular Plants of Ecuador*. Monographs in Systematic Botany, Vol. 75. Missouri Botanical Garden Press, St. Louis.

Kellman, Martin, and Rosanne Tackaberry. 1997. *Tropical Environments. The Functioning and Management of Tropical Ecosystems*. Routledge, London.

Kirkby, A. V. T. 1973. *The Use of Land and Water Resources in the Past and Present Valley of Oaxaca, Mexico*. Memoirs of the Museum of Anthropology, No. 5. University of Michigan, Ann Arbor.

Landon, J. R. (editor). 1991. *Booker Tropical Soil Manual*. Longman Scientific and Technical, Essex, England.

Lathrap, Donald W. 1970. *The Upper Amazon*. Praeger, New York.

Little, Elbert L., and Robert G. Dixon. 1969. *Arboles comunes de la Provincia de Esmeraldas*. Organización de los Naciones Unidas para la Agricultura y Alimentación, Roma.

Maass, J. M. 1995. Conversion of tropical dry forest to pasture and agriculture. Pp. 399–422 in *Seasonally Dry Tropical Forests*. Stephen H. Bullock, Harold A. Mooney, and Ernesto Medina, editors. Cambridge University Press, Cambridge, UK.

Martin, Gary J. 1995. *Ethnobotany. A Methods Manual*. Chapman and Hall, London.

Meggers, Betty J. 1966. *Ecuador*. Praeger, New York.

Moran, Emilio F. 1993. *Through Amazonian Eyes. The Human Ecology of Amazonian Populations*. University of Iowa Press, Iowa City.

Neill, David A., John L. Clark, Homero Vargas, and Tamara Nuñez. 1999. Botanical exploration of the Mache-Chindul mountains, northwestern Ecuador. Final Project Report. Presented to the National Geographic Society, Committee for Research and Exploration.

Newsom, Linda A. 1995. *Life and Death in Early Colonial Ecuador*. University of Oklahoma Press, Norman.

Pearsall, Deborah M. 1980. Analysis of an archaeological maize kernel cache from Manabí Province, Ecuador. *Economic Botany* 34:344–351.

Pearsall, Deborah M. 1994a. Ethnobiological field research. Pp. 145–147 in *Regional Archaeology in Northern Manabí, Ecuador, Vol. 1.*

Environment, Cultural Chronology, and Prehistoric Subsistence in the Jama River Valley. James A. Zeidler and Deborah M. Pearsall, editors. University of Pittsburgh Memoirs in Latin American Archaeology, No. 8. Pittsburgh, PA.

Pearsall, Deborah M. 1994b. Modern agronomic practices and agricultural productivity. Pp. 59–70 in *Regional Archaeology in Northern Manabí, Ecuador, Vol. 1. Environment, Cultural Chronology, and Prehistoric Subsistence in the Jama River Valley*. James A. Zeidler and Deborah M. Pearsall, editors. University of Pittsburgh Memoirs in Latin American Archaeology, No. 8. Pittsburgh, PA.

Pearsall, Deborah M. 1994c. Phytolith analysis. Pp. 161–174 in *Regional Archaeology in Northern Manabí, Ecuador, Vol. 1. Environment, Cultural Chronology, and Prehistoric Subsistence in the Jama River Valley*. James A. Zeidler and Deborah M. Pearsall, editors. University of Pittsburgh Memoirs in Latin American Archaeology, No. 8. Pittsburgh, PA.

Pearsall, Deborah M. 1995a. "Doing" paleoethnobotany in the tropical lowlands: Adaptation and innovation in methodology. Pp. 113–129 in *Archaeology in the Lowland American Tropics. Current Analytical Methods and Recent Applications*. Peter W. Stahl, editor. Cambridge University Press, Cambridge, UK.

Pearsall, Deborah M. 1995b. Domestication and agriculture in the New World tropics. Pp. 157–192 in *Last Hunters—First Farmers: New Perspectives on the Prehistoric Transition to Agriculture*. T. D. Price and A. B. Gebauer, editors. School of American Research Press, Santa Fe, NM.

Pearsall, Deborah M. 1996. Reconstructing subsistence in the lowland tropics: A case study from the Jama River Valley, Manabí, Ecuador. Pp. 233–254 in *Case Studies in Environmental Archaeology*. Elizabeth J. Reitz, Lee A. Newsom, and Sylvia J. Scudder, editors. Plenum Press, New York.

Pearsall, Deborah M. 1999. Agricultural evolution and the emergence of Formative societies in Ecuador. Pp. 161–170 in *Pacific Latin America in Prehistory: The Evolution of Archaic and Formative Cultures*. Michael Blake, editor. Washington State University Press, Pullman.

Pearsall, Deborah M. 2000. *Paleoethobotany. A Handbook of Procedures*. Second Edition. Academic Press, San Diego.

Pearsall, Deborah M., Karol Chandler-Ezell, and Alex Chandler-Ezell. 2003. Identifying maize in neotropical sediments and soils using cob phytoliths. Forthcoming, *Journal of Archaeological Science*.

Pearsall, Deborah M., and John G. Jones. 2001. Prehistoric forest modification in the northern neotropics. Paper presented at the 66th Annual Meeting of the Society for American Archaeology, New Orleans.

Pearsall, Deborah M., and James A. Zeidler. 1994. Regional environment, cultural chronology, and prehistoric subsistence in Northern Manabí. Pp. 201–215 in *Regional Archaeology in Northern Manabí, Ecuador, Vol. 1. Environment, Cultural Chronology, and Prehistoric Subsistence in the Jama River Valley*. James A. Zeidler and Deborah M. Pearsall, editors. University of Pittsburgh Memoirs in Latin American Archaeology, No. 8. Pittsburgh, PA.

Piperno, Dolores R. 1988. *Phytolith Analysis: An Archaeological and Geological Perspective*. Academic Press, Orlando.

Piperno, Dolores R. 1989. The occurrence of phytoliths in the reproductive structures of selected tropical angiosperms and their significance in tropical paleoecology, paleoethnobotany, and systematics. *Review of Palaeobotany and Palynology* 61:147–173.

Piperno, Dolores R., Thomas C. Andres, and Karen E. Stothert. 2000. Phytoliths in *Cucurbita* and other Neotropical Cucurbitaceae and their occurrence in early archaeological sites from the lowland American tropics. *Journal of Archaeological Science* 27:193–208.

Piperno, Dolores R., and Deborah M. Pearsall. 1998. *The Origins of Agriculture in the Lowland Neotropics*. Academic Press, San Diego.

Posey, Darrell A. 1984. A preliminary report on diversified management of tropical forest by the Kayapó Indians of the Brazilian Amazon. *Advances in Economic Botany* 1:112–126. The New York Botanical Garden, Institute of Economic Botany.

Purseglove, J. W. 1968. *Tropical Crops. Dicotyledons*. Longman, Essex, UK.

Purseglove, J. W. 1972. *Tropical Crops. Monocotyledons*. Longman, Essex, UK.

Rees, John D. 1979. Effects of the eruption of Paricutín volcano on landforms, vegetation, and human occupancy. Pp. 249–292 in *Volcanic Activity and Human Ecology*. Payson D. Sheets and Donald K. Grayson, editors. Academic Press, New York.

Rindos, David. 1984. *The Origins of Agriculture: An Evolutionary Perspective*. Academic Press, Orlando.

Sauer, Carl O. 1952. *Agricultural Origins and Dispersals*. American Geographical Society, New York.

Saville, Marshall H. 1907. *The Antiquities of Manabí, Ecuador. A Preliminary Report*. Contributions to South American Archaeology, Vol. I. The George G. Heye Expedition, New York.

Saville, Marshall H. 1910. *The Antiquities of Manabí, Ecuador. Final Report*. Contributions to South American Archaeology, Vol. II. The George G. Heye Expedition, New York.

Schultes, Richard Evans, and Siri von Reis (editors). 1995. *Ethnobotany: Evolution of a Discipline*. Discorides Press, Portland, OR.

Sheets, Payson D. 1983. Summary and conclusions. Pp. 275–293 in *Archaeology and Volcanism in Central America. The Zapotitán Valley of El Salvador*. Payson D. Sheets, editor. University of Texas Press, Austin.

Sheets, Payson D. 1994a. The Proyecto Prehistórico Arenal: An introduction. Pp. 1–23 in *Archaeology, Volcanism, and Remote Sensing in the Arenal Region, Costa Rica*. Payson D. Sheets and Brian R. McKee, editors. University of Texas Press, Austin.

Sheets, Payson D. 1994b. Summary and conclusions. Pp. 312–325 in *Archaeology, Volcanism, and Remote Sensing in the Arenal Region, Costa Rica*. Payson D. Sheets and Brian R. McKee, editors. University of Texas Press, Austin.

Simmonds, N. W. (editor). 1976. *Evolution of Crop Plants*. Longman, London.

Smith, Nigel J. H., J. T. Williams, Donald L. Plucknett, and Jennifer P. Talbot. 1992. *Tropical Forests and Their Crops*. Cornell University Press, Ithaca, NY.

Stahl, Peter W. 2000. Archaeofaunal accumulation, fragmented forests, and anthropogenic landscape mosaics in the tropical lowlands of prehispanic Ecuador. *Latin American Antiquity* 11:241–257.

Thornton, Ian W. B. 2000. The ecology of volcanoes: Recovery and reassembly of living communities. Pp. 1057–1081 in *Encyclopedia of Volcanos*. Haraldur Sigurdsson, Editor-in-Chief. Academic Press, San Diego.

Thurston, H. David. 1997. *Slash/Mulch Systems. Sustainable Methods for Tropical Agriculture*. Westview Press, Boulder, CO.

Timothy, David H., William H. Hatheway, Ulysses J. Grant, Manuel C. Torregroza, Daniel V. Sarria, and Daniel A. Varela. 1963. *Races of Maize in Ecuador*. National Academy of Sciences—National Research Council, Publication 975, Washington DC.

Ugolini, F. C., and R. J. Zasoski. 1979. Soils derived from tephra. Pp. 83–124 in *Volcanic Activity and Human Ecology*. Payson D. Sheets and Donald K. Grayson, editors. Academic Press, New York.

Valencia, Renato, Henrik Balslev, Walter Palacios, David Neill, Carmen Josse, Milton Tirado, and Flemming Skov. 1998. Diversity and family composition of trees in different regions of Ecuador: A sample of 18 one-hectare plots. Pp. 569–584 in *Forest Biodiversity in North, Central, and South America, and the Caribbean. Research and Monitoring*. F. Dallmeier and J. A. Comiskey, editors. Man and the Biosphere Series, Vol. 21. UNESCO, Paris and The Parthenon Publishing Group, Pearl River, NY.

Valverde B., Flor de Maria. 1974. *Claves taxonómicas de ordenes, famílias, y géneros de Dicotiledoneas. I. Parte*. Universidad de Guayaquil, Guayaquil, Ecuador.

van Wambeke, Armand. 1992. *Soils of the Tropics. Properties and Appraisal*. McGraw-Hill, New York.

Veintimilla, Cesar I. 1998. *Analysis of Past Vegetation in the Jama River Valley, Manabí Province, Ecuador*. Masters Thesis, University of Missouri, Columbia.

Veintimilla, Cesar I. 2000. Reconstrucción paleo-ambiental y evolución agrícola en el Valle del Río Jama, Provincia de Manabí, Ecuador. *Revista del Museo Antropológico del Banco Central del Ecuador, Guayaquil* 9:135–151.

Vickers, William T., and Timothy Plowman. 1984. *Useful Plants of the Siona and Secoya Indians of Eastern Ecuador*. *Fieldiana,* Botany, New Series, No. 15. Field Museum of Natural History, Chicago.

Zeidler, James A. 1987. The chroniclers of Coaque and Pasao: Ethnohistorical perspectives on the Jama-Coaque II polity of coastal Ecuador at A.D. 1531. Paper presented at the 86th Annual Meeting of the American Anthropological Association, Chicago.

Zeidler, James A. 1994a. Archaeological testing in the Lower Jama Valley. Pp. 99–109 in *Regional Archaeology in Northern Manabí, Ecuador, Vol. 1. Environment, Cultural Chronology, and Prehistoric Subsistence in the Jama River Valley*. James A. Zeidler and Deborah M. Pearsall, editors. University of Pittsburgh Memoirs in Latin American Archaeology, No. 8. Pittsburgh, PA.

Zeidler, James A. 1994b. Archaeological testing in the Middle Jama Valley. Pp. 71–98 in *Regional Archaeology in Northern Manabí, Ecuador, Vol. 1. Environment, Cultural Chronology, and Prehistoric Subsistence in the Jama River Valley*. James A. Zeidler and Deborah M. Pearsall, editors. University of Pittsburgh Memoirs in Latin American Archaeology, No. 8. Pittsburgh, PA.

Zeidler, James A. 1995. Archaeological survey and site discovery in the forested neotropics. Pp. 7–41 in *Archaeology in the Lowland American Tropics. Current Analytical Methods and Recent Applications*. Peter W. Stahl, editor. Cambridge University Press, Cambridge, UK.

Zeidler, James A., Caitlin E. Buck, and Clifford D. Litton. 1998. Integration of archaeological phase information and radiocarbon results from the Jama River Valley, Ecuador: A Bayesian approach. *Latin American Antiquity* 9:160–179.

Zeidler, James A., and Robin C. Kennedy. 1994. Environmental setting. Pp. 13–41 in *Regional Archaeology in Northern Manabí, Ecuador, Vol. 1. Environment, Cultural Chronology, and Prehistoric Subsistence in the Jama River Valley*. James A. Zeidler and Deborah M. Pearsall, editors. University of Pittsburgh Memoirs in Latin American Archaeology, No. 8. Pittsburgh, PA.

Zeidler, James A., and Deborah M. Pearsall. 1994a. Preface, pp. xvii–xix; The Jama Valley archaeological/paleoethnobotanical project: An introduction. Pp. 1–12 in *Regional Archaeology in Northern Manabí, Ecuador, Vol. 1. Environment, Cultural Chronology, and Prehistoric Subsistence in the Jama River Valley*. James A. Zeidler and Deborah M. Pearsall, editors. University of Pittsburgh Memoirs in Latin American Archaeology, No. 8. Pittsburgh, PA.

Zeidler, James A., and Deborah M. Pearsall (editors). 1994b. *Regional Archaeology in Northern Manabí, Ecuador, Vol. 1. Environment, Cultural Chronology, and Prehistoric Subsistence in the Jama River Valley*. James A. Zeidler and Deborah M. Pearsall, editors. University of Pittsburgh Memoirs in Latin American Archaeology, No. 8. Pittsburgh, PA.

Zeidler, James A., and Marie J. Sutliff. 1994. Definition of ceramic complexes and cultural occupation in the Jama Valley. Pp. 111–130 in *Regional Archaeology in Northern Manabí, Ecuador, Vol. 1. Environment, Cultural Chronology, and Prehistoric Subsistence in the Jama River Valley*. James A. Zeidler and Deborah M. Pearsall, editors. University of Pittsburgh Memoirs in Latin American Archaeology, No. 8. Pittsburgh, PA.

Zimmerman, Laurie S. 1994. Palynological analysis. Pp. 175–183 in *Regional Archaeology in Northern Manabí, Ecuador, Vol. 1. Environment, Cultural Chronology, and Prehistoric Subsistence in the Jama River Valley*. James A. Zeidler and Deborah M. Pearsall, editors. University of Pittsburgh Memoirs in Latin American Archaeology, No. 8. Pittsburgh, PA.

Index